Mechanical and Electro-chemical Interactions Under Tribocorrosion

European Federation of Corrosion Publications
Number 70

Mechanical and Electro-chemical Interactions Under Tribocorrosion

From Measurements to Modeling for Building a Relevant Monitoring Approach

Edited by

Pierre Ponthiaux
CentraleSupelec, Paris-Saclay University,
Gif-sur-Yvette, France

Jean-Pierre Celis
KU Leuven, Leuven, Belgium

(On behalf of WP18 Tribocorrosion)

Published for the European Federation of Corrosion by Woodhead Publishing

Published by Woodhead Publishing Limited on behalf of the European Federation of Corrosion

Woodhead Publishing is an imprint of Elsevier
The Officers' Mess Business Centre, Royston Road, Duxford, CB22 4QH, United Kingdom
50 Hampshire Street, 5th Floor, Cambridge, MA 02139, United States
The Boulevard, Langford Lane, Kidlington, OX5 1GB, United Kingdom

Copyright © 2021 European Federation of Corrosion. Published by Elsevier Ltd. All rights reserved.

No part of this publication may be reproduced or transmitted in any form or by any means, electronic or mechanical, including photocopying, recording, or any information storage and retrieval system, without permission in writing from the publisher. Details on how to seek permission, further information about the Publisher's permissions policies and our arrangements with organizations such as the Copyright Clearance Center and the Copyright Licensing Agency, can be found at our website: www.elsevier.com/permissions.

This book and the individual contributions contained in it are protected under copyright by the Publisher (other than as may be noted herein).

Notices
Knowledge and best practice in this field are constantly changing. As new research and experience broaden our understanding, changes in research methods, professional practices, or medical treatment may become necessary.

Practitioners and researchers must always rely on their own experience and knowledge in evaluating and using any information, methods, compounds, or experiments described herein. In using such information or methods they should be mindful of their own safety and the safety of others, including parties for whom they have a professional responsibility.

To the fullest extent of the law, neither the Publisher nor the authors, contributors, or editors, assume any liability for any injury and/or damage to persons or property as a matter of products liability, negligence or otherwise, or from any use or operation of any methods, products, instructions, or ideas contained in the material herein.

Library of Congress Cataloging-in-Publication Data
A catalog record for this book is available from the Library of Congress

British Library Cataloguing-in-Publication Data
A catalogue record for this book is available from the British Library

ISBN: 978-0-12-823765-6 (print)
ISBN: 978-0-12-823766-3 (online)

For information on all Woodhead publications
visit our website at https://www.elsevier.com/books-and-journals

Publisher: Matthew Deans
Acquisitions Editor: Christina Gifford
Editorial Project Manager: Emily Thomson
Production Project Manager: Vignesh Tamil
Cover Designer: Matthew Limbert

Typeset by STRAIVE, India

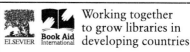

Contents

European Federation of Corrosion publications: Series introduction	ix
Volumes in the EFC series	xi
Contributors	xv

1 Preamble   1
Pierre Ponthiaux and Jean-Pierre Celis

2 Stress/strain effects on electrochemical activity: Metallurgical/mechanical/interactions and surface reactivity   7
Halina Krawiec and Vincent Vignal
   2.1 Introduction   7
   2.2 Definition of a relevant approach   7
   2.3 Influence of compressive stresses   8
   2.4 Influence of tensile stresses in the elastic domain   12
   2.5 Influence of plastic deformation   15
   2.6 Conclusions   23
   2.7 Notations   24
   References   24

3 Mechanical and chemical coupling in tribocorrosion: In situ and ex situ characterization techniques   29
Dominique Thierry, Andrej Nazarov, and Dan Persson
   3.1 Introduction   29
   3.2 Local electrochemical techniques   34
   3.3 Application of in situ vibrational spectroscopy for studies of tribochemical and tribocorrosion processes   54
   3.4 Conclusions   59
   3.5 List of symbols   61
   3.6 List of abbreviations   62
   References   62

4 Managing tribocorrosion investigations by stress mapping: Dual mobility concept, hip implant, as required step   67
Jean Geringer and Caroline Richard
   4.1 Introduction   67
   4.2 Testing of human gait in vitro as an example   72

	4.3	Applications in tribocorrosion	77
	4.4	Conclusions	79
	4.5	Notations	80
	Acknowledgments		80
	References		80
	Further reading		81

5 Parallel wear tests: The need for statistical analysis in tribology — 83
E.P. Georgiou, D. Drees, T. Van der Donck, S. Economou, and Jean-Pierre Celis

5.1	Introduction: Lab testing approach to industrial issues of tribocorrosion	83
5.2	Need for statistics to analyze wear	85
5.3	State-of-the-art on current wear generators	90
5.4	Industrial examples on statistical wear analysis	94
5.5	Summary	108
5.6	Notations	108
	References	108

6 The use of the Pearson's correlation coefficients to identify mechanical-physical-chemical parameters controlling the tribocorrosion of metallic alloy — 111
Vincent Vignal and Halina Krawiec

6.1	Introduction	111
6.2	The Pearson correlation method	112
6.3	Methodology	119
6.4	Application of the PCM in corrosion science and engineering	120
6.5	Application of the PCM to stress corrosion cracking and fatigue corrosion	123
6.6	Application of the PCM to tribocorrosion	124
6.7	Conclusions	128
6.8	Notations	128
	References	129

7 Fretting wear analysis through a mechanical friction energy approach: Impact of contact loadings and ambient conditions — 131
Siegfried Fouvry

7.1	Introduction	131
7.2	Fretting test experiments	132
7.3	Fretting mapping approach	137
7.4	Wear rate description	138
7.5	Influence of contact loadings regarding wear rate fluctuations	145
7.6	Influence of the ambient conditions	150

7.7	Surface wear modeling: Prediction of the maximum wear depth	154
7.8	Conclusion	161
7.9	Notations	163
References		164

8 Harmonic analysis of tribocorrosion: Identification of repassivation kinetics and separation of reactive (corrosion) and mechanical (wear) contributions — 169
Michel Keddam and Vincent Vivier

8.1	Introduction	169
8.2	General concept, experimental approach, and techniques	170
8.3	Analysis of repassivation kinetics by ac modulation of the angular velocity of the counter-body	173
8.4	From steady-state to electrochemical impedance ac modulation of the applied potential	179
8.5	Conclusions and perspectives	187
8.6	Notations	188
References		189
Further reading		189

9 Modeling erosion-corrosion in metals: The effects of elastic rebound and impact angle on erosion-corrosion maps — 191
M.M. Stack and B.D. Jana

9.1	Introduction	191
9.2	Energy balance during a solid particle erosion event	192
9.3	Derivation of an expression for the coefficient of restitution	193
9.4	Derivation of erosion models based on the rebound analysis	195
9.5	Corrosion rate and boundary conditions	198
9.6	Elastic rebound effects on the transition boundaries of the aqueous erosion-corrosion map	200
9.7	Discussion	222
9.8	Conclusions	223
9.9	Notations	224
References		225

10 A monitoring management strategy in tribocorrosion: Application to erosion-corrosion in oil and gas exploration-production — 227
Antoine Surbled

10.1	Introduction	227
10.2	Erosion-corrosion in oil and gas production	228
10.3	Erosion and corrosion management strategy	231
10.4	Conclusions	251
10.5	Notations	253
References		253

Index — 259

European Federation of Corrosion publications: Series introduction

The European Federation of Corrosion (EFC), founded in 1955, is a federation of 41 organizations with interests in corrosion and is based in 24 different countries throughout Europe and beyond. Its member societies represent the corrosion interests of more than 25,000 engineers, scientists, and technicians. The Federation's aim is to advance the science of the corrosion and protection of materials by promoting cooperation in Europe and collaboration internationally. Asides from national and international corrosion societies, universities, research centers and companies can also become Affiliate Members of the EFC.

The administration of the Federation is in the hands of the Board of Administrators (BoA), chaired by the EFC President, and the scientific and technical affairs are the responsibility of the Science and Technology Advisory Committee (STAC), chaired by the STAC Chairman and assisted by the Scientific Secretary. The General Assembly approves any EFC policy prepared and presented by the BoA. The Federation is managed through its General Secretariat with three shared headquarters located in London, Paris, and Frankfurt.

The EFC carries out its most important activities through its more than 20 active working parties devoted to various aspects of corrosion and its prevention, covering a large range of topics including: Corrosion and Scale Inhibition, Corrosion by Hot Gases and Combustion Products, Nuclear Corrosion, Environment Sensitive Fracture, Surface Science and Mechanisms of Corrosion and Protection, Physicochemical Methods of Corrosion Testing, Corrosion Education, Marine Corrosion, Microbial Corrosion, Corrosion of Steel in Concrete, Corrosion in Oil and Gas Production, Coatings, Corrosion in the Refinery Industry, Cathodic Protection, Automotive Corrosion, Tribo-Corrosion, Corrosion of Polymer Materials, Corrosion and Corrosion Protection of Drinking Water Systems, and Corrosion of Archaeological and Historical Artifacts. The EFC is always open to formulating new working parties in response to the demands brought about by developing technologies and their ensuing corrosion requirements and applications.

The European Federation of Corrosion's flagship event is EUROCORR, the most important Corrosion Congresses in Europe, which is held annually in a different European country in September of each year. To date, 27 EUROCORR conferences have taken place in 12 different countries and they have gained a reputation for their high technical quality, global perspective, and enjoyable social program. Another channel for the EFC's valuable transfer of knowledge is the EFC "green" book series which are the fruit of the collaboration and high scientific caliber within and amongst the EFC working party members and are emblematic of the EFC editorial policy.

EFC offices are located at:

European Federation of Corrosion, The Institute of Materials, Minerals and Mining, 1 Carlton House Terrace, London SW1Y 5DB, United Kingdom

Federation Européenne de la Corrosion, Fédération Françhise pour les sciences de la Chimie, 28 rue Saint-Dominique, F-75007 Paris, France

Europäische Föderation Korrosion, DECHEMA e.V., Theodor-Heuss-Allee 25, D-60486 Frankfurt-am-Main, Germany

Volumes in the EFC series

1. **Corrosion in the nuclear industry**
   Prepared by Working Party 4 on Nuclear Corrosion
2. **Practical corrosion principles**
   Prepared by Working Party 7 on Corrosion Education (out of print)
3. **General guidelines for corrosion testing of materials for marine applications**
   Prepared by Working Party 9 on Marine Corrosion
4. **Guidelines on electrochemical corrosion measurements**
   Prepared by Working Party 8 on Physico-Chemical Methods of Corrosion Testing
5. **Illustrated case histories of marine corrosion**
   Prepared by Working Party 9 on Marine Corrosion
6. **Corrosion education manual**
   Prepared by Working Party 7 on Corrosion Education
7. **Corrosion problems related to nuclear waste disposal**
   Prepared by Working Party 4 on Nuclear Corrosion
8. **Microbial corrosion**
   Prepared by Working Party 10 on Microbial Corrosion
9. **Microbiological degradation of materials and methods of protection**
   Prepared by Working Party 10 on Microbial Corrosion
10. **Marine corrosion of stainless steels: chlorination and microbial effects**
    Prepared by Working Party 9 on Marine Corrosion
11. **Corrosion inhibitors**
    Prepared by the Working Party on Inhibitors (out of print)
12. **Modifications of passive films**
    Prepared by Working Party 6 on Surface Science
13. **Predicting CO2 corrosion in the oil and gas industry**
    Prepared by Working Party 13 on Corrosion in Oil and Gas Production (out of print)
14. **Guidelines for methods of testing and research in high temperature corrosion**
    Prepared by Working Party 3 on Corrosion by Hot Gases and Combustion Products
15. **Microbial corrosion: Proceedings of the 3rd International EFC Workshop**
    Prepared by Working Party 10 on Microbial Corrosion
16. **Guidelines on materials requirements for carbon and low alloy steels for H2Scontaining environments in oil and gas production**
    Prepared by Working Party 13 on Corrosion in Oil and Gas Production
17. **Corrosion resistant alloys for oil and gas production: guidance on general requirements and test methods for H2S service**
    Prepared by Working Party 13 on Corrosion in Oil and Gas Production
18. **Stainless steel in concrete: state of the art report**
    Prepared by Working Party 11 on Corrosion of Steel in Concrete
19. **Sea water corrosion of stainless steels: mechanisms and experiences**
    Prepared by Working Party 9 on Marine Corrosion and Working Party 10 on Microbial Corrosion

20. **Organic and inorganic coatings for corrosion prevention: research and experiences**
    Papers from EUROCORR '96
21. **Corrosion_deformation interactions**
    CDI '96 in conjunction with EUROCORR '96
22. **Aspects of microbially induced corrosion**
    Papers from EUROCORR '96 and EFC Working Party 10 on Microbial Corrosion
23. **CO2 corrosion control in oil and gas production: design considerations**
    Prepared by Working Party 13 on Corrosion in Oil and Gas Production
    Published by Woodhead Publishing Limited, 2013
24. **Electrochemical rehabilitation methods for reinforced concrete structures: a state of the art report**
    Prepared by Working Party 11 on Corrosion of Steel in Concrete
25. **Corrosion of reinforcement in concrete: monitoring, prevention and rehabilitation**
    Papers from EUROCORR '97
26. **Advances in corrosion control and materials in oil and gas production**
    Papers from EUROCORR '97 and EUROCORR '98
27. **Cyclic oxidation of high temperature materials**
    Proceedings of an EFC Workshop, Frankfurt/Main, 1999
28. **Electrochemical approach to selected corrosion and corrosion control**
    Papers from the 50th ISE Meeting, Pavia, 1999
29. **Microbial corrosion: proceedings of the 4th International EFC Workshop**
    Prepared by the Working Party on Microbial Corrosion
30. **Survey of literature on crevice corrosion (1979_1998): mechanisms, test methods and results, practical experience, protective measures and monitoring**
    Prepared by F. P. Ijsseling and Working Party 9 on Marine Corrosion
31. **Corrosion of reinforcement in concrete: corrosion mechanisms and corrosion protection**
    Papers from EUROCORR '99 and Working Party 11 on Corrosion of Steel in Concrete
32. **Guidelines for the compilation of corrosion cost data and for the calculation of the life cycle cost of corrosion: a working party report**
    Prepared by Working Party 13 on Corrosion in Oil and Gas Production
33. **Marine corrosion of stainless steels: testing, selection, experience, protection and monitoring**
    Edited by D. Fe´ron on behalf of Working Party 9 on Marine Corrosion
34. **Lifetime modeling of high temperature corrosion processes**
    Proceedings of an EFC Workshop 2001
    Edited by M. Schütze, W.J. Quadakkers, and J.R. Nicholls
35. **Corrosion inhibitors for steel in concrete**
    Prepared by B. Elsener with support from a Task Group of Working Party 11 on Corrosion of Steel in Concrete
    Published by Woodhead Publishing Limited, 2013
36. **Prediction of long term corrosion behavior in nuclear waste systems**
    Edited by D. Fe´ron on behalf of Working Party 4 on Nuclear Corrosion
37. **Test methods for assessing the susceptibility of prestressing steels to hydrogen induced stress corrosion cracking**
    By B. Isecke on behalf of Working Party 11 on Corrosion of Steel in Concrete

38. **Corrosion of reinforcement in concrete: mechanisms, monitoring, inhibitors and rehabilitation techniques**
    Edited by M. Raupach, B. Elsener, R. Polder, and J. Mietz on behalf of Working Party 11 on Corrosion of Steel in Concrete
39. **The use of corrosion inhibitors in oil and gas production**
    Edited by J.W. Palmer, W. Hedges, and J.L. Dawson on behalf of Working Party 13 on Corrosion in Oil and Gas Production
40. **Control of corrosion in cooling waters**
    Edited by J.D. Harston and F. Ropital on behalf of Working Party 15 on Corrosion in the Refinery Industry
41. **Metal dusting, carburization and nitridation**
    Edited by H. Grabke and M. Schütze on behalf of Working Party 3 on Corrosion by Hot Gases and Combustion Products
42. **Corrosion in refineries**
    Edited by J.D. Harston and F. Ropital on behalf of Working Party 15 on Corrosion in the Refinery Industry
43. **The electrochemistry and characteristics of embeddable reference electrodes for concrete**
    Prepared by R. Myrdal on behalf of Working Party 11 on Corrosion of Steel in Concrete
44. **The use of electrochemical scanning tunneling microscopy (EC-STM) in corrosion analysis: reference material and procedural guidelines**
    Prepared by R. Lindström, V. Maurice, L. Klein, and P. Marcus on behalf of Working Party 6 on Surface Science
45. **Local probe techniques for corrosion research**
    Edited by R. Oltra on behalf of Working Party 8 on Physico-Chemical Methods of Corrosion Testing
46. **Amine unit corrosion survey**
    Edited by J.D. Harston and F. Ropital on behalf of Working Party 15 on Corrosion in the Refinery Industry
    Published by Woodhead Publishing Limited, 2013
47. **Novel approaches to the improvement of high temperature corrosion resistance**
    Edited by M. Schütze and W. Quadakkers on behalf of Working Party 3 on Corrosion by Hot Gases and Combustion Products
48. **Corrosion of metallic heritage artefacts: investigation, conservation, and prediction of long term behavior**
    Edited by P. Dillmann, G. Béranger, P. Piccardo, and H. Matthiesen on behalf of Working Party 4 on Nuclear Corrosion
49. **Electrochemistry in light water reactors: reference electrodes, measurement, corrosion, and tribocorrosion**
    Edited by R.-W. Bosch, D. Féron, and J.-P. Celis on behalf of Working Party 4 on Nuclear Corrosion
50. **Corrosion behavior and protection of copper and aluminum alloys in seawater**
    Edited by D. Féron on behalf of Working Party 9 on Marine Corrosion
51. **Corrosion issues in light water reactors: stress corrosion cracking**
    Edited by D. Féron and J.-M. Olive on behalf of Working Party 4 on Nuclear Corrosion
52. **Progress in corrosion: the first 50 years of the EFC**
    Edited by P. McIntyre and J. Vogelsang

53. **Standardization of thermal cycling exposure testing**
    Edited by M. Schütze and M. Malessa on behalf of Working Party 3 on Corrosion by Hot Gases and Combustion Products
54. **Innovative pre-treatment techniques to prevent corrosion of metallic surfaces**
    Edited by L. Fedrizzi, H. Terryn, and A. Simões on behalf of Working Party 14 on Coatings
55. **Corrosion-under-insulation (CUI) guidelines, third ed.**
    Edited by Gino DeLandtsheer on behalf of Working Party 15 on Corrosion in the Refinery and Petrochemical Industry
56. **Corrosion monitoring in nuclear systems**
    Edited by S. Ritter and A. Molander on behalf of Working Party 4 on Nuclear Corrosion
57. **Protective systems for high temperature applications**
    Edited by M. Schütze on behalf of Working Party 3 on Corrosion by Hot Gases and Combustion Products
    Published by Woodhead Publishing Limited, 2013
58. **Self-healing properties of new surface treatments**
    Edited by L. Fedrizzi, W. Fürbeth, and F. Montemor on behalf of Working Party 14 on Coatings
59. **Sulfur-assisted corrosion in nuclear disposal systems**
    Edited by F. Druyts, D. Féron, and B. Kursten on behalf of Working Party 4 on Nuclear Corrosion
60. **Methodology of crevice corrosion testing for stainless steels in natural and treated seawaters**
    Edited by U. Kivisäkk, B. Espelid, and D. Féron on behalf of Working Party 9 on Marine Corrosion
61. **Inter-laboratory study on electrochemical methods for the characterization of CoCrMo biomedical alloys in simulated body fluids**
    Edited by A. Munoz and S. Mischler on behalf of Working Party 18 on Tribo-Corrosion
62. **Testing tribo-corrosion of passivating materials**
    Edited by J.-P. Celis and P. Ponthiaux on behalf of Working Party 18 on Tribo-Corrosion
63. **The corrosion performance of metals for the marine environment**
    Edited by R. Francis and C. Powell on behalf of Working Party 9 on Marine Corrosion
64. **Recommended practices for corrosion management of pipelines**
    Edited by B. Kermani and C. Chevrot on behalf of Working Party 13 on Corrosion in Oil and Gas Production
65. **Corrosion and conservation of cultural heritage metallic artefacts**
    Edited by P. Dillmann, D. Watkinson, E. Angelini, and A. Adriaens on behalf of Working Party 21 on Corrosion of Archeological and Historical Artefacts
66. **Understanding biocorrosing: fundamentals and applications**
    Edited by T. Liengen, D. Féron, R. Basséguy, and I.B. Beech on behalf of Working Party 10 on Microbial Corrosion
67. **Stress corrosion cracking of nickel based alloys in water-cooled nuclear reactors**
    Edited by D. Féron and R. Staehle on behalf of Working Party 4 on Nuclear Corrosion
68. **Engineering tools for corrosion: design and diagnosis**
    Prepared by L. Lazzari
69. **Nuclear corrosion: research, progress and challenges**
    Edited by Stefan Ritter on behalf of Working Party 4 on Nuclear Corrosion
70. **Mechanical and electro-chemical interactions under tribocorrosion**
    Edited by Pierre Ponthiaux and Jean-Pierre Celis on behalf of Working Party 18 Tribo-Corrosion

Contributors

Jean-Pierre Celis KU Leuven, Leuven, Belgium

D. Drees Falex Tribology NV, Rotselaar, Belgium

S. Economou Ecoinnovations, Halandri, Greece

Siegfried Fouvry Ecole Centrale de Lyon, Université de Lyon, CNRS, UMR 5513, Laboratoire de Tribologie et Dynamique des Systèmes, Ecully, France

E.P. Georgiou Dept. Materials Engineering (MTM), KU Leuven, Leuven; Falex Tribology NV, Rotselaar, Belgium

Jean Geringer Université de Lyon, IMT Mines Saint-Etienne, Centre CIS, INSERM SainBioSE U1059, Saint-Etienne, France

B.D. Jana Department of Engineering and Mathematics, Sheffield Hallam University, Sheffield, United Kingdom

Michel Keddam Sorbonne Université, CNRS, UMR 8235, Laboratoire Interfaces et Systèmes Électrochimiques, Paris, France

Halina Krawiec AGH—University of Science and Technology, Faculty of Foundry Engineering, Krakow, Poland

Andrej Nazarov French Corrosion Institute, Brest, France

Dan Persson Rise Kimab, Kista, Sweden

Pierre Ponthiaux CentraleSupelec, Paris-Saclay University, Gif-sur-Yvette, France

Caroline Richard Université de Tours, GREMAN UMR CNRS 7347/Polytech Tours, Tours, France

M.M. Stack Department of Mechanical and Aerospace Engineering, University of Strathclyde, Glasgow, United Kingdom

Antoine Surbled Surbled Antoine Consultants, Paris, France

Dominique Thierry French Corrosion Institute, Brest, France

T. Van der Donck Dept. Materials Engineering (MTM), KU Leuven, Leuven, Belgium

Vincent Vignal Laboratoire Interdisciplinaire Carnot de Bourgogne (ICB), UMR 6303 CNRS—Université de Bourgogne-Franche-Comté, Cedex, France

Vincent Vivier Sorbonne Université, CNRS, UMR 8235, Laboratoire Interfaces et Systèmes Électrochimiques, Paris, France

Preamble

Pierre Ponthiaux[a] and Jean-Pierre Celis[b]
[a]CentraleSupelec, Paris-Saclay University, Gif-sur-Yvette, France, [b]KU Leuven, Leuven, Belgium

Corrosion phenomena can be significantly accelerated by the simultaneous occurrence of a mechanical load onto the surface of a material. In particular, mechanical loading can result in the formation of cracks and surface defects, along with surface strain and stress fields, which lead to faster diffusion of corrosive ions and/or to the destruction of the oxide protective layers and depassivation.

A typical example of this synergism can be found in the chemical industry, in many off-shore applications, but also in dentistry (brackets) and other biomaterial applications (hip and knee prostheses), since the human body is also a source of electrolytes.

The occurrence of tribocorrosion cannot always be easily recognized in field practice. Examples are the accelerated corrosion of steel conveyors exposed to ambient air of high relative humidity, the failure of electrical connectors in the automotive industry, and the erosion wear of turbine blades (Fig. 1.1).

This book follows our handbook on tribocorrosion entitled "Testing of tribocorrosion of passivating materials supporting research and industrial innovation." It complements that handbook and focuses on "Mechanical and (electro)chemical interactions in tribocorrosion: closing the gap between academia and industry for building a relevant monitoring approach."

The transition from measurements carried out in the laboratory to simulation is analyzed in view of recent academic insights. Then, by taking all of these data into account, the actual approaches to build and implement predictive monitoring in tribocorrosion at the application level in the industry are presented.

As with our handbook, this collective work aims to take stock of advances in the field of tribocorrosion. It is the fruit of the large community of European researchers and engineers active in the field of tribocorrosion gathered under the aegis of the European Corrosion Federation (EFC) in its working party n° 18 "Tribocorrosion."

In this context, the various authors and coauthors have chosen to extract in their respective fields of competence, scientific approaches recognized and experimentally validated to address the phenomenon of tribocorrosion by allowing:

- to conduct a reflection on the determination of the number of tests necessary to guarantee the reproducibility of the phenomenon,
- to identify and understand the contribution of the various mechanisms and mechanochemical couplings involved in the wear mechanism,
- to identify the first-order variables among those linked to the mechanical state, materials, or chemical environment, and
- to propose strategies to control and/or extend the lifetime of industrial systems.

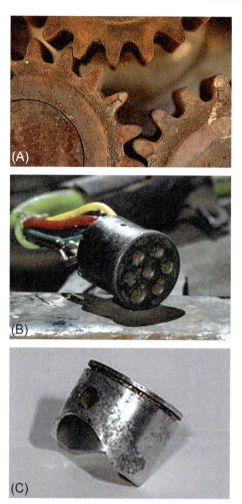

Fig. 1.1 Material degradation in industrial components due to the synergism between corrosion and mechanical stresses: (A) Machinery (https://www.nicepik.com/excavators-stainless-old-construction-machine-rusty-iron-rusted-site-technology-metal-free-photo-295226), (B) Electrical connector (https://www.truckinginfo.com/156050/how-corrosion-causes-electrical-problems), and (C) Car piston (https://www.rcgroups.com/forums/showthread.php?917409-Aluminum-Corrosion-on-OS-piston).

As mentioned in our handbook, solids whose mode of damage or even degradation is worn, belong to the wide range of materials whose microstructure goes from amorphous to crystalline via semicrystalline.

Whatever the type of material used for a technological application, the process for obtaining it and the environment (aerobic or not, degree of humidity, and immersed or not) means that after mechanical stress on the surface, this latter does not have anymore the same structure and composition as the original mass material.

In fact, one cannot, therefore, speak of their surface but a very heterogeneous transition surface and subsurface layer. This transition layer has a variable thickness depending on the method of preparation and uses practice. It is made up of atoms from both the material and the surrounding environment.

On such surfaces, physicochemical phenomena occur, such as oxidation, corrosion, biocorrosion, and embrittlement by hydrogen. They are concomitantly associated with mechanical stresses such as work hardening, cracking, friction or even erosion, abrasion, ... The wear mechanisms resulting from all these loadings and their possible mechanical-(electro)chemical coupling are grouped under the generic term of "Tribocorrosion." That concept appeared in the scientific literature in the early 1980s. It is constructed from the Greek root "tribo" (τριβειν, tribein, rub) and the root of Latin origin "corrodo," the verb which means to gnaw). It was chosen to group under that term, all the phenomena of a physicochemical nature that occur between two surfaces being in contact and relative movement, like work hardening, cracking, and wear, and involve a coupling, synergistic in nature, between mechanical stresses and the environment to which the contact is exposed.

Independently of the classical criteria specific to the field of sliding, like total or partial, continuous or discontinuous, uni or bidirectional, harmonic or random, the degradation mechanisms that occur during fretting-corrosion or fretting wear, slip-corrosion, biotribocorrosion, erosion, abrasion, and cavitation can be analyzed, studied, and even compared. Such studies are mainly related to metallic materials but can be extended to composite materials, ceramics, and other new materials. The modeling of such mechanical-(electro) chemical coupling phenomena requires a necessarily multidisciplinary approach, and can sometimes prove to be delicate and complex due to the appearance of variations of spatial and temporal scales during the establishment of the degradation process and its evolution.

The role of physical (formation of steps, roughness, microcracks), metallurgical (dislocations, crystallographic texture, and grain size), and chemical (passive film composition) changes induced by plastic deformation on electrochemical behavior and corrosion resistance of alloys needs to be considered.

A large part of the tribocorrosion problems can be attributed to a lack of knowledge of the real field conditions existing at the local level which affect the rate of degradation by tribocorrosion of metallic materials used in industrial systems.

On the economic level, the knowledge of the rate of degradation in tribocorrosion of metallic materials represents a major challenge for our society insofar as its knowledge contributes to the control of natural and energetical resources relating to the development of materials, to reliability and the robustness of the sizing of essential structures for the safety of goods and people.

On the environmental level, the information concerning the rate of degradation in tribocorrosion of metallic materials allows in addition to controlling the suitability for the function, taking it into account in a predictive manner and managing the risks of contamination and pollution of soils by metallic trace elements and their evolution according to the physico-chemical conditions of the environment.

To meet all these needs, this book on tribocorrosion presents at first the respective contributions of tests and studies carried out in the laboratory. Their purpose is to identify and understand on the one hand the determination within a tribocorrosion process of the respective weight of the electrochemical, chemical, and mechanical contributions, as well as their possible synergism.

On the other hand, within each of them, to identify the first-order parameters that drive the degradation process(es) to determine their kinetics. The knowledge of the evolution of these parameters as a function of time allows then in particular the development of models which treated numerically make it possible to obtain a behavior simulation for different initial, specific, and field conditions.

An important source of information relies on the application of different local electrochemical techniques to study the electrochemical (corrosion) processes taking place at the worn surface. Examples of such techniques are the scanning vibrating electrode technique, the scanning microcell technique, the scanning Kelvin probe, the Kelvin probe force microscopy, and the scanning electrochemical microscopy which can differently characterize the surfaces at the microresolutions and nanoresolutions. Vibrational spectroscopy introduces spatially resolved chemical information on tribo-films and formed reaction products.

The identification of repassivation kinetics and the separation of reactive (corrosion) and mechanical (wear) degradation in laboratory tests is a major key to unravel the tribocorrosion mechanisms in field practice.

It is also important to acquire information on the local contact stress distribution in specific cases, like the concave-to-concave dual mobility cup system in hip joints. Correlations between surface/subsurface and corrosion parameters are also needed to propose criteria and to understand the basic mechanisms of tribocorrosion.

Recall that a model is a mathematical formalism of a physical system that allows representing of the links (or relation of constraints), existing between given quantities (or variables) of the system. The models used can be of different types and complexity. They can be continuous time or discrete time, qualitative, structural or analytical, linear or nonlinear, representing a correct operational mode or taking failures into account.

In this book, the modeling of the erosion-corrosion in metals is addressed by considering the effect of elastic rebound and impact angle on erosion-corrosion maps, and the predictive monitoring under tribocorrosion conditions applied to the exploitation and the production in the petroleum industry is presented.

Predictive monitoring is mainly based on the estimation of the operating time remaining before a failure occurs, thereby eliminating its suitability for its function. It also serves to define the moment when a maintenance action must be carried out according to a theoretical degradation model of the various mechanical components.

Nevertheless, one of the most difficult industrial issues related to tribocorrosion, is the prediction of long-term wear. Industrial components and products are required to operate for thousands of hours with minimum wear damage.

However, lab-scale tests are accelerated and use high contact pressures and simplified contacts compared to infield applications. Especially in the case of tribocorrosion, the synergism between corrosion and wear complicates the occurring phenomena and further increases the variability of wear. Achieving a high confidence level requires enough data to perform statistical analysis. One way to do that is with the use of parallel testing presented and discussed in this book.

- PROBLEM TREE -

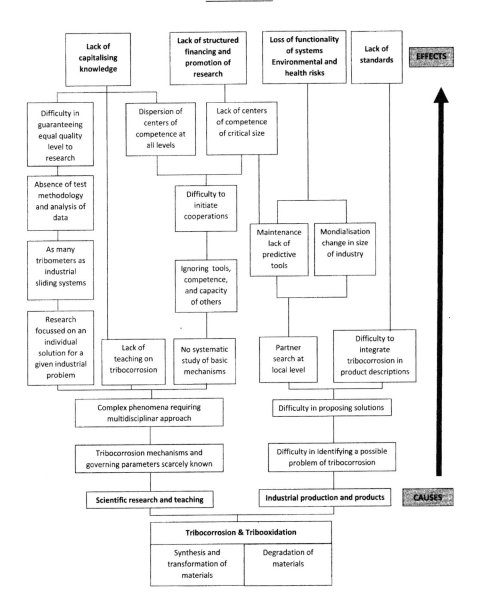

In conclusion, it is obvious that the study of tribocorrosion mechanisms as well as the analyses of the risks of tribocorrosion in industrial systems and their remedies are subjects that require a lot of attention and effort from researchers, designers and maintenance engineers.

This book gives a state-of-the-art on approaches that have been developed or are being developed in academic research centres and in industrial fields.

The "problem tree" analysis shown below gives a pictorial representation of the great diversity of approaches and consequences that are necessary to branch out toward an effective solution when a complex tribocorrosion problem arises.

It helps to find solutions by mapping out the anatomy of cause and effect around an issue of tribocorrosion and tribooxidation. Such a problem tree analysis is best undertaken by brainstorming through information-sharing meetings where a variety of stakeholders are brought together.

Readers of this book and our previous handbook on tribocorrosion will appreciate the usefulness of that "problem tree" for determining the essentials in their case.

Stress/strain effects on electrochemical activity: Metallurgical/mechanical/ interactions and surface reactivity

2

Halina Krawiec[a] and Vincent Vignal[b]
[a]AGH—University of Science and Technology, Faculty of Foundry Engineering, Krakow, Poland, [b]Laboratoire Interdisciplinaire Carnot de Bourgogne (ICB), UMR 6303 CNRS—Université de Bourgogne-Franche-Comté, Cedex, France

2.1 Introduction

Investigating the tribological contact and sliding between two materials is complicated by the fact that the interface is buried from view and inaccessible to conventional experimental tools. In addition, the tribological contact and sliding of materials is a complex and dynamic system impacted by many factors. They involve from macro to the atomic scale, material transfer, heating, and combination of spatially and temporally varying loads. These loads may be triaxial/biaxial/uniaxial (tensile or compressive), or rotational (strength shear). Stresses and strains may develop within the system. The tribological contact and the sliding of materials can also induce phase transformation due to deformation, formation of point and linear defects (voids, cracks, and scratches), and so on.

In an aggressive environment as in tribocorrosion, the physical-mechanical processes mentioned above are affected by corrosion and vice versa. Therefore, the effect of stress and strain on the electrochemical behavior of metallic surfaces needs to be considered and quantified.

The influence of mechanical loads on cathodic (reduction reactions) and anodic processes (dissolution for instance), as well as on the stability and modifications of passive films are discussed in this chapter. The thermokinetic approach is not discussed and readers interested in thermokinetic formalisms are referred to reference [1].

2.2 Definition of a relevant approach

Numerous factors can affect the reactivity of metallic surfaces. Under certain conditions, synergistic effects between two or more factors can also be observed at different scales. Possible factors include:

- Surface roughness (or surface microroughness),
- Crystallographic texture and grain size,
- Chemical composition (alloying elements, for instance),
- Stress/strain. They can be residual fields or applied (external) loads,
- Point and linear defects (scratches, steps, particles, and inclusions),
- Passive films/coatings/contaminants (lubricant and grease).

It is therefore important to first use a statistical approach (Pearson matrix [2], ANOVA module, for example) to determine the parameters whose influence is of order 1 in the given conditions.

Once the most influent parameters are defined, a relevant experimental approach must be selected to study the role of stress/strain. In particular, it must be ensured that the response of the studied system is not significantly affected by factors other than mechanical factors (stress/strain). For example, the influence of surface roughness must be taken into account when studying the role of deformation on reactivity using different surface preparations (mechanical grinding/polishing and electrolytic polishing). Different surface finishes [3] can be used to generate stress/strain fields in alloys like machining, grinding, mechanical polishing, burnishing, and sandblasting. Important is that compressive or tensile stresses can be present on material surfaces.

2.3 Influence of compressive stresses

Experimental techniques used to study the influence of compressive stresses on corrosion (and tribocorrosion) mechanisms will be relevant if they make it possible to generate compressive stresses in a material with negligible work hardening rate, and not inducing changes in (micro)structure and surface roughness. Examples of techniques used to generate compressive stresses are:

- laser shock processing (LSP) [4] applied, e.g., to aluminum alloys [5–7], austenitic and martensitic stainless steels [8], duplex stainless steels [9], nickel alloys [10], and brass [11].
- low plastic burnishing used on AA2024-T3 aluminum alloy [12] and Ca-Mg biomedical alloys [13].

Interested readers are referred to a review paper [14] describing different techniques that can be used to generate compressive stresses. Table 2.1 gives some examples of residual compressive stresses in alloys generated by various techniques.

In all the cases studied, it appears that compressive stresses affect the electrochemical processes and increase the corrosion resistance of metallic alloys. They reduce significantly the kinetics of reduction reactions and the current density in the passive plateau (anodic domain). They can also delay pit initiation and reduce the kinetics of dissolution.

For example, the polarisation behavior of 316L stainless steel in Hank's solution is affected by laser shock peening [15]. Indeed, in the presence of compressive stresses induced by the laser shock (see a red curve in Fig. 2.1), the current density in the cathodic branch of the polarization curve slightly decreases and the current density in the passive range drops by one order of magnitude. In addition, no metastable pits

Table 2.1 Examples of residual compressive stresses induced in metallic alloys using different techniques.

Alloys	Residual stress value (MPa)	Techniques	References
Friction stir welded AA7075	−80	Laser peening	[5]
	−120	Shot peening	
AA2050-T8	−165	Laser peening	[7]
Duplex stainless steel UNS S31803	−155 (γ-phase) −446 (α-phase)	4000 grit surface finish	[9]
Austenitic stainless steel 316L	−257	Rubber polisher #240	[3]
Brass 260	−140	Laser shock processing	[11]
Brass 280	−107		
AA2024-T3	−340	Low plastic burnishing	[12]
Magnesium-calcium implants	[−70; −90]	Low plastic burnishing	[13]

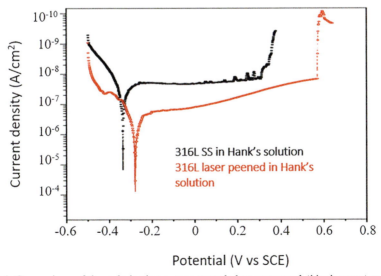

Fig. 2.1 Comparison of the polarisation curves recorded on unpeened *(black curve)* and laser shock peened *(red curve)* 316L stainless steel in Hank's solution.
Based on R.K. Gupta, N. Prasad, A.K. Rai, R. Biswal, R. Sundar, A. Bose, P. Ganesh, K. Ranganathan, K.S. Bindra, R. Kaul, Corrosion study on laser shock peened 316L stainless steel in simulated body fluid and chloride medium, Lasers Manuf. Mater. Process. 5 (2018) 270–282.

are observed after polarisation in the passive range of the laser peened sample. The pitting potential of a laser peened sample shifts to more noble values: a difference of about 150 mV is noticed between unpeened and laser peened samples.

The mechanisms responsible for the improvement in corrosion resistance due to compressive stresses are still under debate. The beneficial influence of compressive stresses cannot be explained considering the density of dopants in the passive films. Indeed, it was shown on 316L stainless steel [16] that the density of dopants (both n- and p-types) is not affected by changing the level of compressive stresses. Some explanations can be found in surface analyses of passive films. Auger measurements were performed after immersion of different stainless steels (302 [17] and 316 [18]) in 1M H_2SO_4 for 20 min. Two samples were considered: a reference sample (no stress) and a sample under compression (generated by bending). Obtained results revealed that the passive film formed in the presence of a compressive stress field is slightly thinner than that formed on the reference surface (thickness of 3.5 nm under compression and 4 nm on the stress-free surface). The passive film formed under compression is also richer in chromium in its inner part than the passive film formed on the reference surface. Fig. 2.2B shows the existence of a peak in the chromium profile of the sample under compression, between 300 and 700 s. Such a peak is not visible on the reference sample (Fig. 2.2A). The protective properties of passive films are usually attributed to the formation of chromium-based compounds in their inner part.

The residual stress field is usually highly heterogeneous in alloys. The relationships between the passive film composition (determined from Auger electron spectroscopy measurements), local residual stresses in the substrate (calculated using a thermal-mechanical simulation based on XRD measurements), and the local electrochemical behavior (via the measurement of the local corrosion potential using microcapillary-based experiments) were studied at the microscale [19]. It was shown that there is a linear relationship between the local corrosion potential and the ratio Cr/Fe in the passive film and between the local corrosion potential and the local residual stress in the

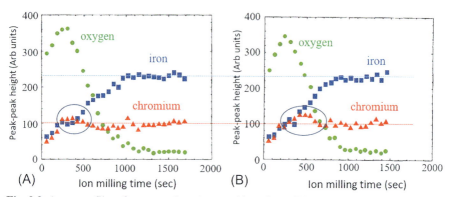

Fig. 2.2 Auger profiles of oxygen, chromium, and iron (ion milling rate of 0.48 nm/min equiv. Si) on: (A) polished sample (no stress, two experiments performed) and (B) sample under compression.
Based on F. Navai, J. Mater. Sci. 30 (1995) 1166–1172.

substrate. The corrosion potential becomes nobler with increasing compressive stresses in the substrate and with increasing Cr/Fe in the passive film. It was then proposed that the passive film formed on sites under compression is enriched in Cr. This study confirms previous those obtained on 302 and 316 stainless steel (previous paragraph).

The influence of an LSP treatment on the polarisation curve of aluminum alloy AA2050-T8 recorded in 0.1M NaCl at 25°C is shown in Fig. 2.3A [7]. After LSP, similar results to those obtained on stainless steels were found. A drop of the current density is observed both in the cathodic and anodic domains. The pitting potential is also shifted to more noble values. AFM observations show that after LSP only pitting corrosion occurs (Fig. 2.3B), whereas both pitting and intergranular corrosion were found after polishing (Fig. 2.3C). Pitting corrosion occurs in sites containing intermetallic particles due to a galvanic coupling between particles and matrix. No clear explanation was proposed to explain changes detected in the corrosion behavior of aluminum alloys.

Corrosion processes and mechanisms are strongly associated with surface electronic structures such as electron work function (EWF). EWF is the minimum work needed to extract electrons from the Fermi level of a metal/alloy across a surface

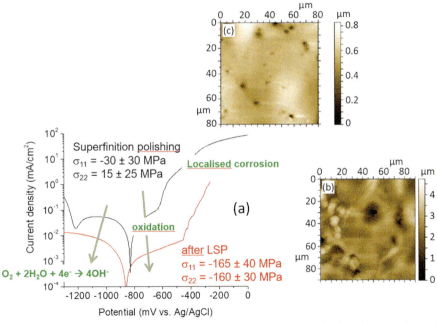

Fig. 2.3 (A) Polarisation curves of AA2050-T8 in 0.1M NaCl at 25°C after either mechanical polishing or LSP. AFM images after corrosion tests (in tapping mode) on (B) LSP treated and (C) mechanical polished aluminium.
Based on H. Amar, V. Vignal, H. Krawiec, C. Josse, P. Peyre, S.N. da Silva, L.F. Dick, Corros. Sci. 53 (2011) 3215–3221.

carrying no net charge. This parameter is highly sensitive to surface conditions (roughness, passive film, stress, strain, and defects). Several papers revealed that EWF is affected by stress/strain [20–25] and a few of these papers [24, 25] showed that compressive stress increases the EWF. The higher the EWF value, the more stable the surface and therefore the higher the corrosion resistance. Recent progress in understanding the corrosion behavior of metallic alloys using SKP can be found in reference [26]. Regarding tribocorrosion, it was underlined that some surface properties, such as adhesion and friction, are largely influenced by surface electron activity, which can be reflected by EWF [27]. Relevant experiments [28] have demonstrated a close correlation between EWF and adhesive force for surfaces in contact.

The beneficial effect of compressive stresses on the corrosion behavior of metallic materials is however not a general rule. Indeed that may be the case when other facts appear simultaneously with the development of compressive stresses like the formation of both a rough surface and a cold worked layer (plastic deformation) near the surface as in the case of alloy 800 [29] and the decohesion of the surface layer in the case of carbon steel [30].

2.4 Influence of tensile stresses in the elastic domain

The transition between domains where compressive and tensile stresses exist is very important since, at that transition, the corrosion resistance of metallic alloys decreases. This is the case for duplex stainless steels in sodium chloride solutions [9] The time to a pit of UNS S31803 duplex stainless steel was measured vs. applied stress (located in the elastic domain), as shown in Fig. 2.4. Experiments were performed in 0.5M NaCl at 50°C, under potentiostatic conditions (at 300 and 400 mV vs SCE). Surface stresses were also measured by means of XRD in both phases. Two domains are visible. In domain I, eight experiments (of the 10 experiments performed) leads to passive behavior (no time to pit identified within 100 h). By contrast, only three experiments (of the 10 tests performed) leads to passive behavior in domain II. At the transition between the two domains, a drop in the corrosion resistance is observed. Residual stress

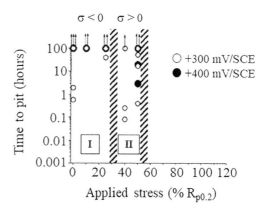

Fig. 2.4 Evolution of the time to pit vs applied stress for UNS S31803 duplex stainless steel in 0.5M NaCl at 50°C, under potentiostatic conditions. Based on V. Vignal, N. Mary, C. Valot, R. Oltra, L. Coudreuse, Electrochem. Solid-State Lett. 7 (2004) C39–C42.

measurements (using XRD) showed that this transition corresponds to the transition from compressive to tensile stresses in the austenite phase (the ferrite phase being under compression in both domains).

The deleterious influence of tensile stresses on the electrochemical behavior and corrosion resistance was also reported for the following metallic alloys, namely 316L stainless steel [16], ferritic steel [31], 304L stainless steel [32, 33], high-strength galvanized steel [34], carbon steel [35], X80 steel [36], aluminum alloys 7075 [37] and 7010 [38], Fe-based amorphous coatings [39], and 316L stainless steel [40]. Table 2.2 gives some examples of tensile residual stresses generated in metallic alloys using various techniques.

The experimental techniques used to induce tensile stresses in the elastic domain are relevant when they generate tensile stresses in the material with a negligible work hardening rate, and with no change neither in (micro) structure or surface roughness. This is illustrated in Fig. 2.5 showing the effect of stresses induced by mechanical

Table 2.2 Examples of residual tensile stresses induced in alloys using different techniques.

Metals and alloys	Residual stress value (MPa)	Techniques	References
High-strength structural steel	+190	Tempering at 100°C (30 min)	[41]
5A12 aluminum alloy	[+50; +110]	Nd: YAG laser cleaning (50–110 W)	[42]
High strength alloy steel	[+600; +800]	Machining	[43]
40Cr	+70	Ultrahigh speed grinding	[44]
Austenitic stainless steel 316L	+19	Electropolishing	[3]
	+315	Angle grinder	
X65 steel (weld zone)	+160	Welding	[45]

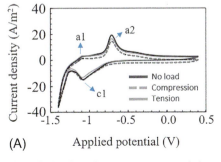
(A)

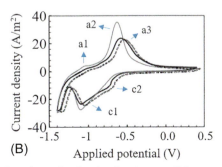
(B)

Fig. 2.5 Cyclic voltammograms recorded on C-ring electrodes (made of carbon steel) in (A) a NaCl-free solution, and (B) a solution containing 0.85M NaCl. Based on Y. Zhang, A. Poursaee, J. Mater. Civil Eng. 17 (8) (2015).

loadings at different levels on the cyclic voltammograms recorded on carbon steel immersed in simulated concrete pore solutions containing either or not chloride ions [35]. The peaks observed on these voltammograms correspond to the oxidation or reduction of compounds present in the passive film, namely:

- Peak a1 is attributed to the formation of $Fe(OH)_2$ and FeO
- Peak a2 is attributed to the reaction $Fe^{2+} \rightarrow Fe^{3+}$
- Peak a3 is attributed to an oxidation within the compact passive film (formation of Fe_2O_3, Fe_3O_4 and/or FeOOH)
- Peak c1 is related to the reduction reactions associated with the anodic peak a2
- Peak c2 is attributed to reduction reactions associated with the anodic peak a3

In the absence of chloride ions in the solution (Fig. 2.5A), there is no significant influence of either compressive or tensile stresses on the electrochemical response of the system. By contrast, large differences in electrochemical response were observed in the presence of chloride ions on carbon steel exhibiting tensile stresses in the elastic domain (Fig. 2.5B). For example, the current density at peak a2 (oxidation peak) is significantly increased. One reason could be the fact that chloride ions preferentially adsorb onto metallic surfaces in the presence of tensile stresses [46]. That passive film is then preferentially perturbed at locations where tensile stresses are present.

In literature, the deleterious influence of tensile stresses in the elastic domain was often attributed to changes in the electrical conductivity of passive films. The density of dopants increases significantly at increasing tensile stresses, as shown in Fig. 2.6 [40]. A linear relationship between the density of dopants and the level of tensile stress (in the elastic domain, was found for 316L stainless steel in alkaline solutions. The inner layer (density of acceptors) of the passive film in contact with the substrate is more affected by tensile stresses than the outer layer.

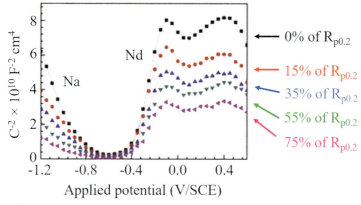

Fig. 2.6 Influence of tensile stress in the elastic domain on the Mott-Schottky plots of 316L stainless steel in $0.05M\ H_3BO_3 + 0.075M\ Na_2B_4O_7 \cdot 10H_2O$ (buffered solution, pH 9.2). N_a is the density of acceptors and N_d is the density of donors.
Based on L. Jinlong, L. Tongxiang, W. Chen, G. Ting, Mater. Sci. Eng. C 61 (2016) 32–36.

Table 2.3 Galvanic corrosion potentials and current densities of AA5050 vs 40CrNiMoA steel.

	Galvanic corrosion potentials (mV vs SCE)		Average current density ($\mu A/cm^2$)
Stress	Aluminum alloy	Steel	
No stress	−775	−653	31.7
25%$R_{p0.2}$ Al	−890	−654	43.95
50%$R_{p0.2}$ Al	−912	−639	50.86
75%$R_{p0.2}$ Al	−923	−636	58.16

Based on L. Jinlong, L. Tongxiang, W. Chen and G. Ting, Mater. Sci. Eng. C 61 (2016) 32–36.

The evolution of the kinetic constants, k_0, of an oxidation reaction (mediator oxidation) at the 304L surface, was quantified by scanning electrochemical microscopy for various tensile stresses in the elastic domain [33]. k_0 significantly increased with increasing tensile stress, indicating that the reaction is promoted. According to the Point Defect Model [47], this phenomenon was ascribed to the acceleration of the transport of oxygen vacancies. This analysis is consistent with the experimental results obtained from the Mott-Schottky plots presented above.

Tensile stresses in the elastic domain may also affect the galvanic coupling between two metallic alloys. So, e.g., when aluminum alloy (AA5050) undergoes tensile stress after being coupled to 40CrNiMoA steel in 3.5% NaCl at 30°C for 20 h, corrosion was promoted [48]. With increasing stress, the potential of AA5050 decreases whereas the potential of steel remains more-or-less unchanged (Table 2.3). The potential difference between this aluminum alloy and 40CrNiMo steel (driving force for the galvanic coupling) then increases and thus promotes the galvanic coupling between the two alloys, ending up in the anodic dissolution of AA5050 (less noble alloy). The average current density (calculated over the 20 h of immersion) increases as the stress applied to the aluminum alloy increases (Table 2.3). It is increased by a factor of 2 at 75%$R_{p0.2}$ when to compared to the stress-free situation.

2.5 Influence of plastic deformation

Plastic deformation can be achieved by applying external loads like, e.g., uniaxial tensile tests, bending tests, extrusion tests, but also by surface preparation methods like, e.g., mechanical polishing, by elaboration methods, e.g., rolling, and by processing methods like, e.g., welding and cutting techniques. Applying loads in the plastic domain of the stress-strain curves may induce several changes in a metallic alloy, namely:

- Formation of steps at the surface resulting from the emergence of slip bands,
- Activation of dislocations (both geometrically necessary (GND) and statistically stored dislocations (SSD)). GND accommodates nonuniform strains at the crystal scale. They are associated with the local incompatibility of the crystal lattice due to geometrical constraints. By contrast, SSD form by the random trappings of dislocations. Note that only GND induces changes of the lattice orientation and can be detected using diffraction-based techniques [49]. A large amount of GND may accumulate in the material to form geometrically necessary dislocation boundaries. Then SSD are stored between these geometrically necessary dislocation boundaries,
- Increase of the surface roughness,
- Grain refinement,
- Formation of microcracks, microvoids,
- Changes in the passive film properties (composition, structure, thickness).

These numerous changes can affect the electrochemical parameters in the plastic domain. In addition, synergistic effects between two or more of these parameters may occur and some parameters may have antagonistic effects. Therefore, the study of the influence of plastic deformation on the electrochemical response of metallic alloys is a difficult task. It is hard to propose quantitative relationships between plastic deformation and electrochemical parameters. The reason is that nonlinear evolutions are often found. In general plastic, deformation has a deleterious influence on the electrochemical processes and the corrosion resistance of metallic alloys [50–55]. The evolution of the amount of metastable pits (Fig. 2.7A) and electrical charge (Fig. 2.7B) vs. cold rolling reduction is shown in Fig. 2.7 [55]. In both cases, nonlinear relationships were found. It is worth mentioning that different laws were obtained for the two parameters. The amount of metastable pits increases with increasing cold rolling reduction up to 20% and then decreases. By contrast, the electrical charge increases continuously with the cold rolling reduction.

The effect of plastic deformation on the corrosion processes in aqueous media can be linked to geometrical effects, crevice effects, dislocations effects, and modifications of the passive film [55].

2.5.1 Geometrical effects due to the presence of slip bands

Numerous alloys have a crystallographic orientation-dependent corrosion behavior [56]. In this case, local techniques in physics and electrochemistry are powerful tools to investigate corrosion mechanisms. For example, Electron backscatter diffraction (EBSD) analyses can be coupled to local electrochemical measurements using microcapillary-based techniques. These techniques allow the selection and interrogation of single sites and are therefore becoming increasingly popular to investigate the electrochemical behavior of metallic grains/phases and nonmetallic heterogeneities in alloys. This methodology was applied to pure aluminum [57] to determine the corrosion behavior of individual grains whose crystallographic orientation was previously determined by means of EBSD.

It was revealed that pitting corrosion occurs after polishing along grains oriented along (111) planes. There is no pitting corrosion on grains oriented along (001) and (101) planes. For example, Fig. 2.8A shows polarisation curves obtained in several

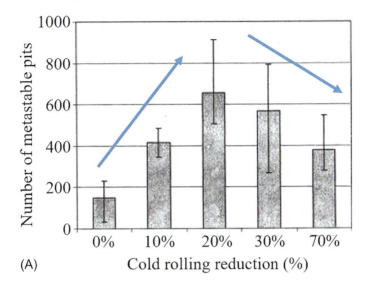
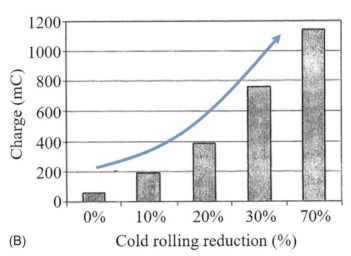

Fig. 2.7 Evolution of (A) the number of metastable pits and (B) electrical charge vs cold rolling reduction measured on 304 immersed at open circuit potential in a solution containing 0.1M NaCl + 2 × 10^{-4} M FeCl$_3$.
Based on L. Peguet, B. Malki, B. Baroux, Corros. Sci. 49 (2007) 1933–1948.

(001) grains. In this case, a large passive range is observed and no pitting potential can be defined. The same behavior was obtained on (101) grains.

By contrast, pitting corrosion occurs on grains oriented along (001), (101), and (111) planes after plastic deformation of 5.5% (Fig. 2.8B showing polarisation curves in different (001) grains). This behavior was also observed on grains possessing a very low density of dislocations. It was proposed that pits initiate at steps generated by the

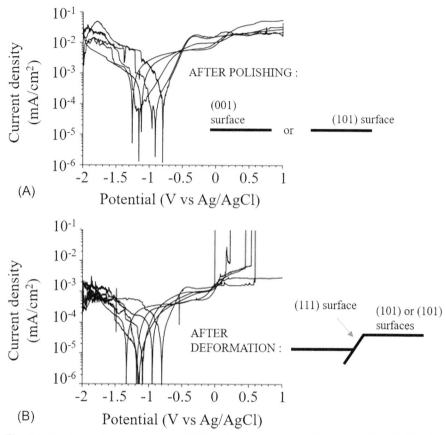

Fig. 2.8 Local polarisation curves of (001) grains of pure aluminium in 0.1M NaCl (A) after polishing and (B) a plastic deformation of 5.5%.
Based on H. Krawiec, Z. Szklarz, Electrochim. Acta 203 (2016) 426–438.

emergence of slip bands at a material surface. Indeed, these steps are oriented along (111) planes (cubic system). This orientation exhibits the lowest corrosion resistance after polishing. Pitting potentials measured after deformation on (001) and (101) grains are located within the same potential range as those measured on (111) oriented grains after polishing, confirming the previous assumption.

2.5.2 Crevice effect due to the presence of cracks

Microcracks and microvoids can be generated at material surfaces during plastic deformation. Depending on the geometry of these defects, a crevice effect may occur in aggressive environments. The presence of deep and narrow microcracks in AlCu4Mg1 alloy after plastic deformation induces a drop of the pitting potential in NaCl media [58]. The physical-chemical conditions in the microcracks promote the initiation and propagation of stable pits. A similar effect was observed in metallic

Stress/strain effects on electrochemical activity

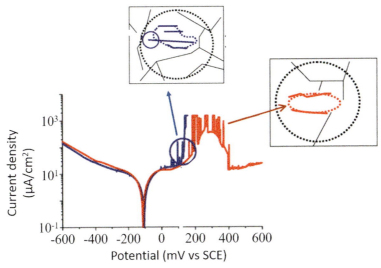

Fig. 2.9 Influence of the morphology of microcracks generated during plastic deformation (6%) on the local polarisation curves of 316L stainless steel containing MnS inclusions in 1M NaCl [61]. Based on Oltra R., Vignal V., Corros. Sci. 49 (2007) 158-165.

alloys containing MnS-inclusions after plastic deformation [59–62]. The transition from metastable to stable pitting depends on the location of microcracks [61]. Stable pits are systematically observed in presence of microcracks located in the matrix (see a blue curve in Fig. 2.9). By contrast, only metastable pits were found when microcracks propagated along with the interface between matrix and inclusions (see a red curve in Fig. 2.9). The finite element method can be used to predict the location where microcracks initiate. The initiation depends on the mechanical properties of both the matrix and the inclusions, and the morphology of inclusions.

2.5.3 Modifications of the passive film

Changes in the electrochemical behavior of Ti6Al4V in the Ringer's solution at 37°C were associated with modifications of the chemical composition of the passive film [63]. Auger measurements revealed that the beneficial influence of the presence of a mixed passive film composed of aluminum and titanium compounds is lowered after straining in the plastic domain (5%). The decrease of the quantity of aluminum in the passive film is associated with an increase of the current density in the passive range.

Changes in the structure of the passive film formed on pure iron [64], Ti6Al4Valloy [65], and 304 stainless steel [66] were also considered to explain the influence of plastic deformation in the increase of corrosion susceptibility. The adsorption energy of oxygen on a Cu (100) surface affects the build-up of Cu_2O oxide films dependent on the stress applied to the substrate [67].

2.5.4 Presence of dislocations

During plastic deformation, GND and SSD are generated and dislocation structures are formed. The characteristics of dislocation structures depend on the deformation process and materials [68]. The density of dislocations can be calculated and the dislocations structures identified using transmission electron microscopy (TEM), for example. EBSD can also be used to locate regions with a density of GND. The presence of dislocation decreases the EWF value and therefore increases the corrosion susceptibility of metallic alloys [25].

Coupling local electrochemical measurements and the microstrain gauge technique, it was shown [69] that GND generated at some grain boundaries to accommodate incompatibilities play an important role in the kinetics of cathodic reduction reactions occurring on as-cast AlMg2 alloy after plastic deformation.

The kinetics of the hydrogen reduction reaction but also the recombination reactions can affect when the substrate is subjected to plastic deformation [70]. Such significant effects have been reported on single crystals and polycrystals of nickel in an H_2SO_4 environment. That clearly shows the competing effects associated with the expression of the deformation like dislocation density, stress, and the emergence of deformation bands.

Dislocations play also a crucial role in the initiation of stable pits in various metallic alloys immersed in NaCl-based media like, e.g., pearlitic alloys [71], duplex stainless steels [72]. In the case of pearlitic alloys in 0.05M NaCl (Fig. 2.10), a critical value of the grain orientation spread can be defined, leading to a drop of the pitting potential. Regarding Ta-4%W alloy after plastic deformation, pitting corrosion occurs at regions of original grain boundaries and dislocation boundaries (GNB) [73].

The influence of dislocations on the dissolution of polycrystalline nickel was studied in acidic environment [74, 75]. The thermodynamic approach showed that they could be the reason for the increase of the dissolution rate of the metal by modifying the adsorption sites for hydroxide species.

2.5.5 Influence of grain refinement generated by severe plastic deformation

Interested readers are referred to several review papers [76–78] describing the influence of grain refinement by severe plastic deformation (SPD) on corrosion. It is important to keep in mind that changes observed in the corrosion behavior of alloys with fine/ultrafine grains are imparted by not only grain refinement but also other microstructural changes (absence of porosity, residual stresses, crystallographic orientation, and dislocation density).

A review of the literature [78] indicates that it seems likely that grain refinement (with other microstructural changes induced during refinement) does not compromise the overall corrosion resistance, and in many cases, improve it in comparison with coarse-grained alloys.

The impact of the presence of fine/ultrafine grains on the corrosion behavior of the main families of alloys can be classified as follows (in descending order of impact):

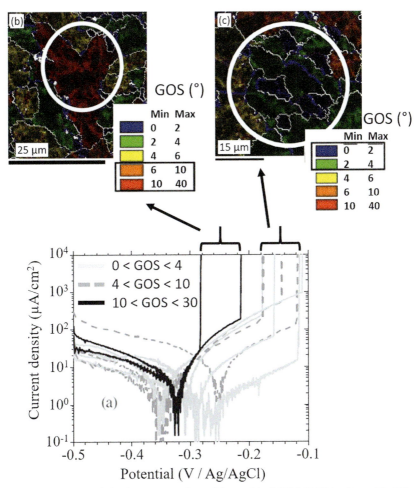

Fig. 2.10 (A) Local polarisation curves of pearlitic steel in 0.05M NaCl in sites with different GOS values. (B and C) Sites where electrochemical analysis were performed [71].
Based on Rault V., Vignal V., Krawiec H., Dufour F., Corros. Sci. 100 (2015) 667-671.

stainless steels, followed by aluminum and magnesium alloys [78]. Table 2.4 reports some examples of corrosion tendency observed in some iron and aluminum alloys after grain refinement. In this Table, the corrosion behavior of the alloy with fine/ultrafine structure is compared to that of the same alloy with coarse structure: The sign "+" means better corrosion resistance of the fine structure whereas the sign "−" means worst corrosion resistance. The sign "=" is used when no significant changes are observed between the two structures. The impact is relatively marginal in pure copper and titanium alloys.

Processing involved in grain refinement can alter the bulk (including the surface) or only the surface of the alloy [78]. Grain refinement can be obtained throughout the

Table 2.4 Examples of corrosion behavior of alloys with fine/ultra-fine structure.

Alloys	Grain size	Corrosion tendency	Techniques	References
Al 7150	120 nm	=	High pressure torsion	[79]
Al 2024	200 nm	+	Equal channel angular pressing	[80]
AZ91D	[1; 2] μm	−	Equal channel angular pressing	[81]
AZ31B	523 nm	+	Ultrasonic peening	[82]
AISI316	20–70 nm	−	Hydraulic extrusion	[83]
Fe-20% Cr	200 nm	+	Equal channel angular pressing	[84]
AISI321	15 nm	=	Ultrasonic peening	[85]

volume of the alloy using different techniques, namely equal channel angular pressing (ECAP), high-pressure torsion (HPT), and accumulative roll bonding (ARB), for example. Grain refinement can also be obtained within a sublayer near the surface using various techniques like high power shot blast, ultrasonic peening, surface mechanical attrition treatment (SMAT), burnishing/brushing (under certain conditions), and friction stir processing. In this case, the volume properties are not altered by processing.

Depending on the alloy and experimental conditions used to refine the structure (type of technique and experimental conditions), the grain size ranges from the microscale (fine structures) down to the nanoscale (ultrafine structures). Table 2.4 gives some examples of grain size values.

Regarding tribocorrosion, grain refinement could enhance the wear resistance of materials by increasing the hardness [76, 86]. It could also enhance the tribocorrosion resistance of alloys mainly by increasing surface hardness and forming protective and lubricating surface layers [86]. Interested readers are referred to a review paper [86] describing the tribocorrosion behavior of nanocrystalline metals. In this paper, nanocrystalline metals are elaborated by means of surface deposition techniques. However, obtained results can be very useful to interpret the corrosion behavior of alloys after grain refinement by severe plastic deformation, as both materials (elaborated by surface deposition techniques and by severe plastic deformation) have comparable grain size. For example, it was noticed on nanocrystalline metals that under tribocorrosion conditions, the presence of a nanostructure could decrease or increase the corrosive rates (depending on the property of the passive and the nature of corrosion products).

Only a few papers investigate the influence of grain refinement (induced by severe plastic deformation) on the tribocorrosion of metals and alloys. This is the case of 304L stainless steel [87] and titanium [88]. In the latter case [87], a nanocrystallized surface layer of around 150-μm thickness was created by surface peening. Electron back-scattered diffraction (EBSD) has put in evidence an interesting point: the

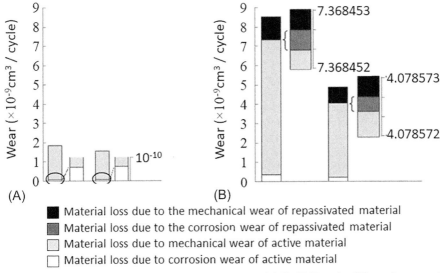

Fig. 2.11 Contribution of wears obtained on nanopeened AISI 304L under (A) continuous and (B) intermittent sliding against alumina pin (5 N, 0.0628 m/s) in filtrate olive pit/tap water. Based on F. Ben Saada, Z. Antar, K. Elleuch, P. Ponthiaux, N. Gey, Wear 394–395 (2018) 71–79.

nanocrystalline surface layer was heterogeneous and composed of two sublayers. The upper layer consists of nanosized ferrite grains whereas the bottom layer was composed of martensite and deformed austenite.

Fig. 2.10 compares the corrosion and mechanical contributions of the total wear loss for the untreated and treated samples (304L stainless steel) [87]. Experiments were performed under continuous (Fig. 2.11A) and intermittent (Fig. 2.11B) sliding. For both sliding conditions and both samples, the mechanical wear contribution was predominant in the total wear loss. Tribocorrosion mechanism was dominated by abrasion mechanical removal of the uncovered surface.

In addition, the total wear loss measured on the treated sample is significantly lower than that of the untreated sample under intermittent sliding (Fig. 2.10B). Under continuous sliding, the difference is very large. This shows that that the treated 304L was more sensitive to tribocorrosion under intermittent sliding than continuous one due to depassivation/repassivation phenomena. This also shows that the difference observed between both samples depends on sliding conditions.

2.6 Conclusions

There is an abundant bibliography on the interactions between mechanical stress/strain and the reactivity of metal alloys. The following conclusions can be drawn:

(1) Regarding the influence of compressive stresses, almost all publications show a positive impact on the electrochemical response and corrosion resistance of alloys. Some elements of response were provided for stainless steels considering the properties of passive films, namely changes in chemical composition.
(2) All publications show that tensile stresses in the elastic domain of stress-strain curves have a deleterious influence on the corrosion resistance of alloys. This is attributed to an increase in the conductivity of passive films.

Several explanations were proposed to explain the influence of compressive and tensile stresses on the electrochemical response of metallic alloys, mainly changes in the electron work function and changes in the properties (composition and structure) of passive films.
(3) Effects obtained in the plastic domain of the stress-strain curves are very difficult to analyze. They depend on many parameters like the activation of several types of dislocations, the nature of the structure of dislocations, the emergence of slip bands, the surface roughness, and the presence of cracks. Nonlinear relationships between plastic deformation and electrochemical parameters are generally found.

2.7 Notations

k_0 kinetic constant of a reaction (m/s)
σ normal stress (Pa)
σ_{11} normal stress along the x_1-axis (Pa)
σ_{22} normal stress along the x_2-axis (Pa)
$R_{p0.2}$ apparent yield strength (Pa)
N_a density of acceptors in the passive film (m^{-2})
N_d density of donors in the passive film (m^{-2})

References

[1] V. Vignal, O. Devos, X. Feaugas, Effect of stress/strain fields on electrochemical activity: metallurgy/stress interaction and surface reactivity, in: C. Blanc, I. Aubert (Eds.), Mechanics—Microstructure—Corrosion Coupling: Concepts, Experiments, Modeling and Cases, ISTE Press Ltd and Elsevier Ltd, London/Kidlington Oxford, 2019, pp. 271–284.
[2] K. Pearson, Philos. Mag. 50 (1900) 157–175.
[3] O. Takakuwa, H. Soyama, Adv. Chem. Eng. Sci. 5 (2015) 62–71.
[4] P. Peyre, V. Vignal, Techniques de l'Ingénieur 1 (2012). COR 1 580.
[5] O. Hatamleh, P.M. Singh, H. Garmestani, Corros. Sci. 51 (2009) 135–143.
[6] U. Trdan, J. Grum, Investigation of corrosion behaviour of aluminium alloy subjected to laser shock peening without a protective coating, Adv. Mater. Sci. Eng. (2015). paper ID 705306.
[7] H. Amar, V. Vignal, H. Krawiec, C. Josse, P. Peyre, S.N. da Silva, L.F. Dick, Corros. Sci. 53 (2011) 3215–3221.
[8] P. Peyre, C. Carboni, Ann. Chim.–Sci. Mat. 29 (2004) 83–93.
[9] V. Vignal, N. Mary, C. Valot, R. Oltra, L. Coudreuse, Electrochem. Solid-State Lett. 7 (2004) C39–C42.
[10] T. Sano, A. Hirose, Rev. Laser Eng. 36 (2008) 91–92.
[11] N. Lisenko, C.D. Evans, Y.L. Yao, Manuf. Lett. 23 (2020) 5–8.
[12] X. Liu, G.S. Frankel, Corros. Sci. 48 (2006) 3309–3329.

[13] M. Salahshoor, Y.B. Guo, Surface integrity of biodegradable magnesium-calcium (Mg-Ca) alloy by low plasticity burnishing, in: Proc. of the STLE/ASME 2010 International Joint Tribology Conference, Paper No. IJTC2010-4121, 2010, pp. 61–63.
[14] S. Kanchidurai, P.A. Krishanan, K. Baskar, K. Saravana, R. Mohan, A review of inducing compressive residual stress—shot peening; on structural metal and welded connection, IOP Conf. Ser. Earth Environ. Sci. 80 (2017). paper 012033.
[15] R.K. Gupta, N. Prasad, A.K. Rai, R. Biswal, R. Sundar, A. Bose, P. Ganesh, K. Ranganathan, K.S. Bindra, R. Kaul, Corrosion study on laser shock peened 316L stainless steel in simulated body fluid and Chloride medium, Lasers Manuf. Mater. Process. 5 (2018) 270–282.
[16] V. Vignal, C. Valot, R. Oltra, M. Verneau, L. Coudreuse, Corros. Sci. 44 (2002) 1477–1496.
[17] F. Navai, J. Mater. Sci. 30 (1995) 1166–1172.
[18] F. Navai, O. Debbouz, J. Mater. Sci. 34 (1999) 1073–1079.
[19] V. Vignal, O. Delrue, O. Heintz, J. Peultier, Electrochim. Acta 55 (2010) 7118–7125.
[20] W. Li, M. Cai, Y. Wang, S. Yu, Scr. Mater. 54 (2006) 921–924.
[21] L. Chen, C. Shi, X. Li, Z. Mi, D. Wang, H. Liu, L. Qiao, Materials 10 (2017). paper 273.
[22] N. Fuertes Casals, A. Nazarov, F. Vucko, R. Pettersson, D. Thierry, J. Electrochem. Soc. 162 (2015) C465–C472.
[23] W. Li, D.Y. Li, J. Phys. D. Appl. Phys. 37 (2004) 948–951.
[24] W. Li, Y. Wang, D.Y. Li, Phys. Status Solidi 201 (2004) 2005–2012.
[25] W. Li, D.Y. Li, Mater. Sci. Technol. 18 (2002) 1057–1060.
[26] B. Łosiewicz, M. Popczyk, M. Szklarska, A. Smołka, P. Osak, A. Budniok, Solid State Phenom. 228 (2015) 369–382.
[27] S. Liu, H. Lu, D.Y. Li, Appl. Surf. Sci. 351 (2015) 316–319.
[28] Y. Li, D.Y. Li, Appl. Phys. Lett. 95 (2004) 7961–7965.
[29] R.K. Zhu, J.L. Luo, Electrochem. Commun. 12 (2010) 1752–1755.
[30] X. Feng, Y. Zuo, Y. Tang, X. Zhao, X. Lu, Electrochim. Acta 58 (2011) 258–263.
[31] S.J. Kim, Int. J. Hydrog. Energy 42 (2017) 19367–19375.
[32] K.R. Trethewey, M. Paton, Mater. Lett. 58 (2004) 3381–3384.
[33] D. Sidane, O. Devos, M. Puiggali, M. Touzet, B. Tribollet, V. Vivier, Electrochem. Commun. 13 (2011) 1361–1364.
[34] W.J. Yang, P. Yang, X.M. Li, W.L. Feng, Mater. Corros. 63 (2012) 401–407.
[35] Y. Zhang and A. Poursaee, J. Mater. Civil Eng. 2015, 27, n°8.
[36] J. Zhang, J. Liu, Q.H. Huang, Z.Y. Cheng, J.T. Guo, Anti-Corros. Method Mater. 62 (2015) 103–108.
[37] M. Dollah, M.J. Robinson, Corros. Eng. Sci. Technol. 46 (2011) 42–48.
[38] P.J.E. Forsyth, Mater. Lett. 41 (1999) 173–180.
[39] Y. Wang, K.Y. Lib, F. Scenini, J. Jiao, S.J. Qu, Q. Luo, J. Shen, Surf. Coat. Technol. 302 (2016) 27–38.
[40] L. Jinlong, L. Tongxiang, W. Chen, G. Ting, Mater. Sci. Eng. C 61 (2016) 32–36.
[41] W. Ding, Y. Liu, J. Xie, L. Sun, T. Liu, F. Yuan, J. Pan, Metals 9 (2019) 709.
[42] G. Zhu, S. Wang, W. Cheng, G. Wang, W. Liu, Y. Ren, Coatings 9 (2019) 578.
[43] H. Jiang, L. He, Z. Ren, F. Shao, S. Yuan, Int. J. Adv. Manuf. Technol. 106 (2020) 4693–4705.
[44] J. Chen, Q. Fang, L. Zhang, Int. J. Adv. Manuf. Technol. 75 (2014) 615–627.
[45] L. Bai, K. Jiang, L. Gao, Mater. Res. 21 (2018), e20180166.
[46] V. Vignal, H. Zhang, O. Delrue, O. Heintz, I. Popa, J. Peultier, Corros. Sci. 53 (2011) 894–903.

[47] D.D. Macdonald, Russ. J. Electrochem. 48 (2012) 235–258.
[48] T. Cui, D. Liu, P. Shi, J. Liu, Y. Yi, H. Zhou, J. Wuhan Univ. Technol. Mater. Sci. Ed. 33 (2018) 688–696.
[49] T. Leffers, T. Lorentzen, The plastic regime, including anisotropy effects, in: M.T. Hutchings, A.D. Krawitz (Eds.), Measurement of Residual and Applied Stress Using Neutron Diffraction, NATO ASI Series, Series E. Applied Sciences 216, 1992, pp. 171–187.
[50] A.H. Ramirez, C.H. Ramirez, I. Costa, J. Braz. Chem. Soc. 25 (2014) 1270–1274.
[51] V. Tandon, A.P. Patil, R.C. Rathod, S. Shukla, Mater. Res. Express 5 (2018), 026528.
[52] B. Mazza, P. Pedeferri, D. Sinigaglia, A. Cigada, G.A. Mondora, G. Re, G. Taccani, D. Wenger, J. Electrochem. Soc. 126 (1979) 2075–2081.
[53] H.B. Li, Z.H. Jiang, Q.F. Ma, Z. Li, Adv. Mater. Res. 217-218 (2011) 1180–1184.
[54] M. Rifai, H. Miyamoto, H. Fujiwara, Mater. Sci. Appl. 5 (2014) 568–578.
[55] L. Peguet, B. Malki, B. Baroux, Corros. Sci. 49 (2007) 1933–1948.
[56] T.Q. Ansari, J.L. Luo, S.Q. Shi, Mater. Degrad. 3 (2019) 1–12.
[57] H. Krawiec, Z. Szklarz, Electrochim. Acta 203 (2016) 426–438.
[58] H. Krawiec, V. Vignal, Z. Szklarz, J. Solid State Electrochem. 13 (2009) 1181–1191.
[59] N. Shimahashia, I. Muto, Y. Sugawara, N. Hara, J. Electrochem. Soc. 161 (2014) C494–C500.
[60] E.G. Webb, T. Suter, R.C. Alkire, J. Electrochem. Soc. 148 (2001) B186–B195.
[61] R. Oltra, V. Vignal, Corros. Sci. 49 (2007) 158–165.
[62] Z. Szklarska-smialowska, Corrosion 28 (1972) 388–396.
[63] H. Krawiec, V. Vignal, J. Loch, P. Erazmus-Vignal, Corros. Sci. 96 (2015) 160–170.
[64] T. Yamamotoz, K. Fushimi, S. Miura, H. Konno, J. Electrochem. Soc. 157 (2010) C231–C237.
[65] D. Nakhaie, A. Davoodi, G.R. Ebrahimi, Corrosion 72 (2016) 110–118.
[66] T. Souier, M. Chiesa, Effect of surface conditions and strain hardening on the passivity breakdown of 304 stainless steel, J. Mater. Res. 27 (2012) 1580–1588.
[67] D. Kramer, Y. Wang, J. Wharton, Faraday Discuss. 180 (2015) 137–149.
[68] D.A. Hughes, N. Hansen, D.J. Bammann, Scr. Mater. 48 (2003) 147–153.
[69] H. Krawiec, V. Vignal, Z. Szklarz, Corros. Sci. 65 (2012) 387–396.
[70] H. El Alami, J. Creus, X. Feaugas, Electrochim. Acta 51 (2006) 4716–4727.
[71] V. Rault, V. Vignal, H. Krawiec, F. Dufour, Corros. Sci. 100 (2015) 667–671.
[72] V. Vignal, D. Ba, H. Zhang, F. Herbst, S. Le Manchet, Corros. Sci. 68 (2013) 275–278.
[73] G. Ma, Q. He, X. Luo, G. Wu, Q. Chen, Materials 12 (2019). paper 117.
[74] M. Sahal, J. Creux, R. Sabot, X. Feaugas, Acta Mater. 54 (2006) 2157–2167.
[75] M. Sahal, C. Savall, J. Creus, R. Sabot, X. Feaugas, Dislocations effect on kinetic of passivation of polycrystalline nickel in H2SO4 medium, in: P. Marcus, V. Maurice (Eds.), Passivation of Metals and Semiconductors, and Properties of Thin Oxide Layers, Elsevier, 2006, pp. 519–524.
[76] K.D. Ralston, N. Birbilis, Corrosion (2010) 66. 075005-075005-13.
[77] E. Karakulak, J. Magnes. Alloy 7 (2019) 355–369.
[78] H. Miyamoto, Mater. Trans. 57 (2016) 559–572.
[79] K.S. Ghosh, N. Gao, M.J. Starink, Mater. Sci. Eng. A 552 (2012) 164–171.
[80] J.G. Brunner, N. Birbilis, K.D. Ralston, S. Virtanen, Corros. Sci. 57 (2012) 209–214.
[81] D. Song, A.B. Ma, J.H. Jiang, P.H. Lin, D.H. Yang, J.F. Fan, Corros. Sci. 53 (2011) 362–373.
[82] Z. Pu, G.L. Song, S. Yang, J.C. Outeiro, O.W. Dillon Jr., D.A. Puleo, I.S. Jawahir, Corros. Sci. 57 (2012) 192–201.

[83] M. Pisarek, P. Kedzierzawski, M. Janik-Czachor, K.J. Kurzydlowski, Corrosion 64 (2008) 131–137.
[84] M. Rifai, H. Miyamoto, H. Fujiwara, Int. J. Corros. 386865 (2015) 1–9.
[85] B.N. Mordyuk, G.I. Prokopenko, M.A. Vasylyev, M.O. Iemov, Mater. Sci. Eng. A 458 (2007) 253–261.
[86] Z. Wang, Y. Yan, L. Qiao, Mater. Trans. 56 (2015) 1759–1763.
[87] F. Ben Saada, Z. Antar, K. Elleuch, P. Ponthiaux, N. Gey, Wear 394-395 (2018) 71–79.
[88] S. Faghihi, D. Li, J.A. Szpunar, Nanotechnology 21 (2010). 485703.

Mechanical and chemical coupling in tribocorrosion: In situ and ex situ characterization techniques

Dominique Thierry[a], Andrej Nazarov[a], and Dan Persson[b]
[a]French Corrosion Institute, Brest, France, [b]Rise Kimab, Kista, Sweden

3.1 Introduction

Tribocorrosion is defined as the simultaneous action of wear and corrosion on moving surfaces in contact [1]. As an example, a passive surface film may be damaged mechanically by wear (e.g., rubbing or impacts). This causes local thinning and failure of the passive film, resulting in the exposure of the bare metal surface (see Fig. 3.1) [2]. Normally, the film will spontaneously reform through metal oxidation, often accompanied by the dissolution of the metal into the environment. This leads to an increase in the corrosion rate. The rate of mechanical surface activation is a function of tribological parameters: mechanical loading, the presence of lubricants, hardness, roughness, and wear debris. Archard's law gives the volume V_m of material removed due to rubbing [3]:

$$V_m = K_w F_N L / H \tag{3.1}$$

where K_w is the wear coefficient, F_N is the applied normal force, L is the sliding distance, and H is the hardness of the material. The theoretical interpretation of the wear coefficient K_w and its numerical value depends on the predominant material removal mechanism, which can be adhesive, abrasive, oxidative, or through fatigue (see Fig. 3.1). The rate of repassivation depends on the corrosive media, e.g., the electrochemical potential of the metal, temperature, pH, or the presence of different oxidizers and inhibitors. In an electrochemical experiment, the amount of anodically oxidized metal is determined from the measured current using Faraday's law:

$$V_{ch} = QM / nF\rho \tag{3.2}$$

where V_{ch} is the volume of metal transformed by anodic dissolution; Q is the electric charge passed, which is obtained by integrating the measured current with respect to time; M is the atomic mass of the metal; n is the charge number for the oxidation reaction (apparent valence); F is the Faraday constant; ρ is the density of the metal.

The schematic for the mechanical activation of the surface is shown in Fig. 3.1 [2]. The indenter mechanically removes material from the worn surface and creates an electrochemically active surface that accelerates the corrosion of the material.

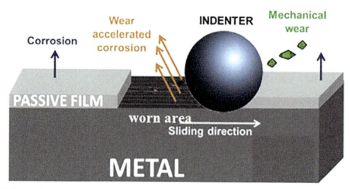

Fig. 3.1 Schematic of the worn surface [2].

The mechanisms of tribocorrosion are complicated, and the rate depends on many mechanical and electrochemical factors. Thus, the electrochemical potential of the metal is influenced not only by the rate of electrochemical dissolution but also by changes in the friction coefficient between the two surfaces, which greatly influences the rate of mechanical wearing. It is supposed that maximal static friction (adhesion, hardness, surface tension) exists at the potential of zero charges and is accompanied by an increase in the damage to the surface [4]. However, applying a suitable potential can reduce the friction in driven devices and result in considerable savings of fuel energy [4, 5].

In corrosion-wear environments, material failures are usually accelerated by synergistic interactions between the mechanical action of wear/rubbing and the corrosion reactions occurring on the worn surfaces (Fig. 3.1). This synergism and acceleration in the metal lost is a result of the mechanical activation of passivity, which leads to an acceleration in metal dissolution on newly formed surfaces. For this reason, the rate of tribocorrosion is influenced by the nature of the electrolyte and the presence of passive films. Thus, for stainless steel and other passive materials, the metal lost in deionized water is the result of abrasion without synergism with corrosion. In aggressive aqueous electrolytes, the presence of NaCl can accelerate tribocorrosion [6, 7]. On the other hand, for active metal surfaces, mechanical activation through sliding has only a small impact on the corrosion rate [8]. Thus, the mechanisms of depassivation and repassivation are important to tribocorrosion.

Fig. 3.2 shows the setups of a classic electrochemical experiment in tribocorrosion [9, 10]. The open-circuit potential (OCP) is monitored using a reversible reference electrode (RE) to display the potential changes due to removing the passive film from the surface by rubbing it with an alumina ball (Fig. 3.2A and B). At the end of the rubbing, the potential increases and returns to its initial level due to the repassivation of the worn surface. The measurement of the corrosion potential during tribocorrosion is a rather simple technique that allows information to be gathered about the surface state of the sliding metal. However, this technique does not provide information on the kinetics of the corrosion reactions.

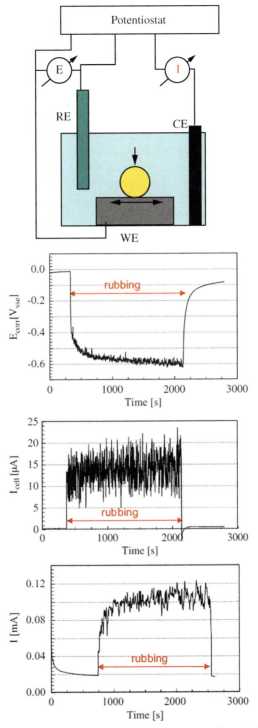

Fig. 3.2 Schematic of macroscale electrochemical measurements for rubbing Ti6Al4V against alumina ball in 0.9% NaCl aqueous electrolyte (A), monitoring of open circuit potential using calomel reference electrode (B), measurements of galvanic currents using counter electrode (cathode) (C) and monitoring of the external current at a fixed potential (D) [9, 10].

Fig. 3.2C shows the monitoring of current using a zero ammeter (I) between working (WE) and counter (CE) electrodes prepared from the same material. The surface of the WE is activated through sliding or rubbing by an alumina ball, which decreases the potential and increases the current. The current returns to zero at the end of the tribocorrosion action due to the repassivation of the worn surface. Fig. 3.2D shows the potentiostatic tribocorrosion experiment using a reference (RE) and counter (CE) electrode. The abrasion of the working electrode (WE) surface damages the passive oxide film, resulting in an anodic current related to the dissolution of the metal from the active locations. At the end of the experiment, the current returns to zero due to the repassivation of the metal surface (Fig. 3.2D). The worn area is small relative to the exposed surface of the working electrode, which means the real current density at the active location is in the range of 5 mA/cm^2 [10].

In the setup shown in Fig. 3.3, the two electrodes are short-circuited through a zero-resistance ammeter that equalizes their potentials. The current related to the difference in the kinetics of the electrochemical reactions on the damaged (worn) and passive (unworn) surfaces flow through the ammeter. The corrosion current density is thus precisely related to the wearing area, as the WE is partially polymer-coated.

Classic electrochemical experiments include the measurement of polarization curves (Fig. 3.4). A comparison of the polarization curves for abraded and nonabraded stainless steel surfaces shows that wearing is the main cause of increases in the rate of anodic dissolution of the steel (Fig. 3.4).

Electrochemical impedance spectroscopy (EIS) can provide information on the conditions of the corroding metal surface [13–17]. As a result, the capacitance and resistance of the surface oxide film, the rate of the activation-passivation process, and the polarization resistance to corrosion can be obtained in rubbing and steady-state conditions [15]. With EIS, the degradation of the passive layer due to fretting-corrosion can be assessed [16].

The corrosion current due to wear (Figs. 3.2 and 3.4) is averaged across all working electrode surfaces. The physical interpretation of the corrosion mechanism requires more precise knowledge of the local state of the worn surface. Fig. 3.5 shows that wear creates dislocations, which result in dislocation pile-ups, residual stress, and microcracks [2]. Mechanical stress leads to refinement of the grains and hardening of the substrate. To gain new insights on tribocorrosion in passive metals, single reaction

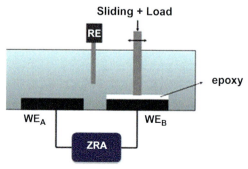

Fig. 3.3 Set-up to study interaction and effect of the wearing. Experimental technique for quantifying the coupling effects on stainless steel during tribocorrosion under equilibrium conditions [11].

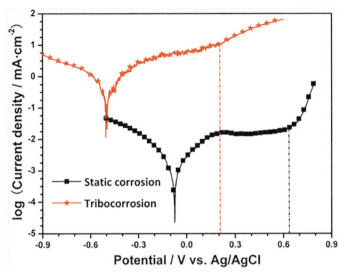

Fig. 3.4 Polarization behavior of S31254 steel with and without sliding wear in seawater [12].

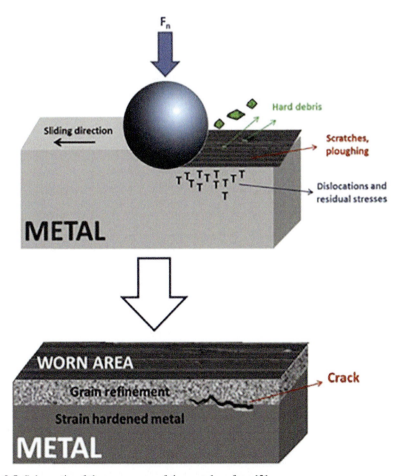

Fig. 3.5 Schematic of the worn area of the metal surface [2].

steps such as local activation and repassivation have to be studied individually, taking into account differences in activity at different locations. Over the past few decades, local electrochemical techniques have been developed and now allow the study of electrochemical behavior as a function of metallurgical factors, such as phase heterogeneity, microstructure, differences in composition, grain size, dislocations, and inclusions. The main objective of this article is to review the work performed using local electrochemical techniques for the study of tribocorrosion at the microlevel and nanolevel.

3.2 Local electrochemical techniques

3.2.1 Scanning reference electrode technique (SRET) and scanning vibrating electrode technique (SVET)

3.2.1.1 General description

In SRET, a reference electrode (e.g., a wire from an inert metal or a capillary from a reversible reference microelectrode) is scanned at 100–200 μm above the surface. This electrode measures the potential as a function of the position. Thus, SRET measures the microgalvanic potential difference, which is local to the surface of the material under investigation. The distribution of local currents can be determined by considering the ionic conductivity of the electrolyte. This nonintrusive technique provides dynamic information on corrosion activity by measuring variations in current flow in the electrolyte at the microscopic scale. Kok et al. [18] used a modified SRET method to investigate localized corrosion of worn surfaces produced during tribocorrosion.

Fig. 3.6 shows linear scans across a stainless steel surface during wear with and without a physical vapor-deposited coating (C/Cr) and with different applied loads. It has been demonstrated that SRET provides quantitative information regarding the separation of the local anodic and cathodic locations during tribocorrosion. The local corrosion currents were smaller for the coated surface compared with the bare

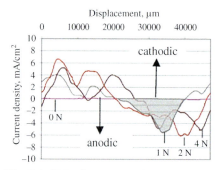
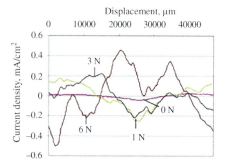

Fig. 3.6 SRET line scans under various applied loads for identical contact times for uncoated stainless steel surface (left) and steel coated by physical vapor deposition (right) [18].

stainless steel. SRET can thus provide the following parameters at the microscale: local corrosion activity, the number of active anodic areas, sample recovery time (repassivation), and open circuit potential [18].

The purpose of SVET is the same as that of SRET, i.e., to measure the corrosion current distribution above a corroding electrode surface in an electrolyte [19–22]. In the case of SVET, the probe is vibrating above the surface; due to vibrations and the integration of the current, the probe is more sensitive to low corrosion currents flowing close to the electrode surface.

Fig. 3.7 shows a schematic of a corroding surface with separated anodic and cathodic locations. The ionic current in the electrolyte is flowing between the local anodes and cathodes. The microprobe (10–30 µm in diameter), which is made of Pt or Pt-Ir, measures the potential drop in a small volume of solution (see Fig. 3.7B). When vibrating in a plane perpendicular to the surface, the tip crosses the electric field generated by the ionic current flowing between the anodic and cathodic locations. After the determination of the ionic conductivity of the electrolyte, the potential drop is converted to a local current flowing at a distance of a few hundred microns above the sample.

3.2.1.2 Scanning vibrating electrode technique. Experimental results for surface abrasion

SVET has been applied to the map pitting corrosion of steel, MnS inclusions [20], cut edge corrosion [19, 22], corrosion of welds [21], and the influence of grain boundaries on pitting corrosion and cracking [23]. Fig. 3.8 shows the distribution of corrosion currents above highly heterogenic surfaces of abrasive alloys with hard particles (W-C-Co alloy) that were embedded into a stainless steel matrix.

The corrosion behavior of hard-metal composites with WC hard particles and AISI 304L stainless steel as binder phases was analyzed. These WC-based systems offer

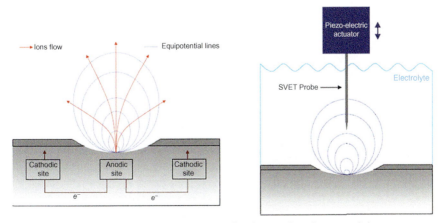

Fig. 3.7 Schematic of the SVET measurements for the metal surface containing electrochemical heterogeneity (A) and schematic of the SVET probe (B).

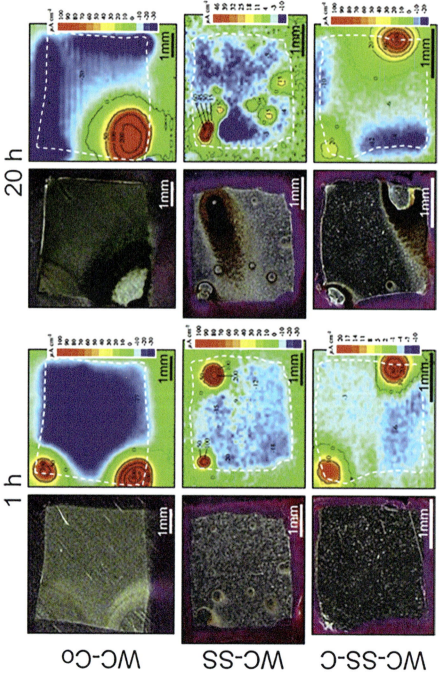

Fig. 3.8 Optical image and SVET current density for three materials after 1 and 20 h of immersion in an aqueous electrolyte, namely, 0.005 M NaCl [24].

improved wear resistance because of their hard ceramic phase surrounded by a metallic binder matrix. The corrosion of the composites took place through the uniform dissolution of the metallic binder phase, together with small pits scattered across the surface. The active anodic locations (pits marked in red on the maps in Fig. 3.8) on the stainless steel matrix were visualized. Large areas with negative current are related to cathodic areas of oxygen reduction, which indicates that the surfaces of all the samples were electrochemically active. During exposure, some pits passivated, and the lowest current was measured for WC-SS composite (Fig. 3.8). However, some abrasive composites showed corrosion activity for a long time [24].

3.2.2 Electrochemical microcell (EMC)

A triboelectrochemical microcell was developed to separately investigate the activation and repassivation at local sites under rubbing conditions and to evaluate the repassivation kinetics [25, 26]. A classical electrochemical setup similar to that shown in Fig. 3.2 was used. However, due to the small dimensions of the cell, the measurements could be localized to smaller areas (e.g., to an individual grain or a single-phase). The local corrosion current density could be measured for a metal probe in an electrolyte under rubbing conditions. A ceramic tube, placed inside a capillary, was used as a rubbing partner. The rotational speed, load, and applied electrochemical potential could be varied. Furthermore, this method allows separate studies of the activation and repassivation of metal surfaces. The principle of the method is shown in Fig. 3.9. An

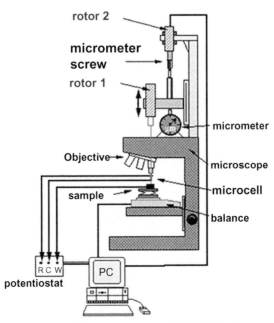

Fig. 3.9 Setup for microtribocorrosion measurements [25, 26]. A microcapillary is used as electrochemical cell, with an inside ceramic tube as a rubbing partner.

increase in the current of dissolution was found at the locations of the MnS inclusions [26]. It was concluded that, in many systems, the susceptibility of a passive metal to tribocorrosion may be related to the removal rate of the protective oxide film and the subsequent repassivation rate of the bare metal exposed to an aggressive environment [26].

For tribocorrosion experiments, the microcell, which consists of a glass microcapillary filled with electrolyte, was applied to a small surface area on the worn surface of some duplex steel [27]. The tip of the capillary (50 μm in diameter) was sealed onto the specimen surface with silicone rubber. The microcell was mounted on a microscope for precise positioning of the capillary on the surface and the entire setup was placed in a Faraday cage. A three-electrode setup made it possible to investigate the distribution of the electrochemical properties at the microlevel. The electrochemical microcell technique was combined with an X-ray microdiffraction technique to determine the influence of surface stress on the electrochemical behavior of the different phases in the duplex steel inside and outside the wear track. Polarization curves were measured at the microscale for the austenite and ferrite phases after sliding. The results show that sliding generates elastic straining of the ferrite and plastic deformation of the austenite. Compressive and shear forces applied during the sliding result in plastic deformation of the austenite and elastic deformation of the ferrite. Thus, the electrochemical behavior of these phases was dramatically altered, inducing a reduction in the corrosion resistance and in the passivation properties [27].

One interesting application of the microcapillary cell is the investigation of activation-passivation processes in a single grain of iron [28]. It was shown that the passivation rate is a function of the surface energy and crystallographic orientation of the grain.

3.2.3 Scanning Kelvin probe

3.2.3.1 Principles of the technique

A study by Stratmann and Streckel [29] introduced SKP to electrochemical corrosion studies. Over the past few years, the method has been applied to the characterization of stress and hydrogen in steel. In contrast to the previous local electrochemical techniques, the probing of the steel surface is carried out in the air. Thus, the method can be extremely useful for analyzing the effect of stress on metals under atmospheric corrosion conditions [30–33].

SKP consists of two metallic electrodes separated by an air gap (Fig. 3.10). Mechanical vibrations of the microprobe above the working electrode create an alternating capacitor that leads to the flow of an alternating current (Eq. 3.3).

$$I(\tau) = \Delta V_{p/w}(dC_{p/w}/dt) \qquad (3.3)$$

$$\Delta V_{p/w} = (\Phi_w - \Phi_P)/e \qquad (3.4)$$

where $I(\tau)$ is the alternating current; $C_{p/w}$ is the capacitance; $\Delta V_{p/w}$ is the contact potential difference between the probe and the working electrode; e is the electron

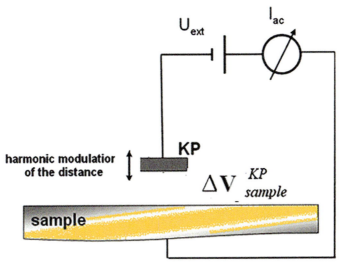

Fig. 3.10 Simplified sketch of SKP measurements under atmospheric conditions [29]. The reference electrode vibrates perpendicular to the sample plane. U_{ext} and I_{ac} the source of backup voltage for compensation of the contact potential difference and the microammeter correspondingly.

charge; Φ_w and Φ_P are the electron work functions of the working electrode and the probe surfaces correspondingly. The potential drop $\Delta V_{p/w}$ is located within the air gap and relates to the difference in Volta potentials or corresponding electron work functions (Eq. 3.4). The Volta potential of the probe (Φ_P/e) remains constant and calibrated against any reference electrode. Thus, the potential (Φ_w/e) or work function (Φ_w) of the working electrode surface can be obtained.

Φ is defined as the minimum work to extract an electron from the Fermi level (μ_e) and transfer it to a point just outside the metal surface (carrying no net charge). Φ_w consists of two parts: the chemical potential of an electron inside the metal (μ_e) and the potential drop (X_w) across the metal/air interface (Eq. 3.5).

$$X_w = (\mu_{ox} - \mu_e)/e + F_b + \beta_{ox/air} \tag{3.5}$$

The first part is the metal/oxide contact potential difference, which is proportional to the difference in Fermi levels between electrons in the metal (μ_e) and the oxide (μ_{ox}). e is the elementary electron charge. The second potential drop (F_b) relates to the adsorption of environmental components (molecules of oxygen and water), which bends the conductive and valence bands in semiconducting oxide films. The electric charge in the oxide is counterbalanced by the charges of adsorbate species (e.g., O_2^- ions) creating a potential drop ($\beta_{ox/air}$). Thus, the Kelvin probe can measure the local electronic properties of the surface oxide under the impact of corrosion, stress, or hydrogen flux [29–34]. The spatial resolution of modern SKP instruments is in the range of 50 μm. The topography and potential distribution are measured simultaneously. On the other hand, atomic force microscopy in the SKP model (KPFM) can reach better spatial resolutions and will be discussed later.

3.2.3.2 SKP application to tribocorrosion and experimental results

Tribocorrosion applies a mechanical force that creates dislocations, pile-ups of dislocations, built-in residual stress, and microcracks on a worn surface (Fig. 3.5). Hence, SKP could be a valuable tool for studying tribocorrosion [30–34]. According to Eq. (3.5), changes in the state of the surface oxide (passive) film will result in changes to the electrochemical potential of the metal. Nazarov et al. demonstrate that different surface defects (MnS inclusions, scratches, and stress concentrations in bends) on a surface of 304 stainless steel can be visualized using SKP [30].

Fig. 3.11 shows that the scratching of stainless steel and high strength steel decreases the surface potential, which results in lower passivity. Soft stainless steel shows a more significant decrease compared to harder carbon steel. Low-potential locations can be prone to localized corrosion attacks.

Wear creates dislocations and residual stress. It was found that the SKP technique can visualize stress distributions in austenitic stainless steel (Fig. 3.12). The application of plastic strain creates dislocations and dislocation pile-ups (Fig. 3.12B). The emergence of dislocations from the bulk to the steel surface makes the passive oxide film more defective and decreases the Volta potential around the deformation (Fig. 3.12C). The measurements were carried out in situ under load and at the rest after unloading the sample with residual stress.

The surface of the 301 LN steel in the notch decreased the potential by 250 mV after the application of a 10% tensile strain (Fig. 3.12C). The photograph shows an increase in the surface roughness due to the emergence of dislocations around the notch with residual stress [32]. Thus, SKP can identify locations with elevated stress. More negative locations are related to steel surfaces with lower levels of passivity. These locations are more prone to pitting corrosion and stress corrosion cracking.

Fig. 3.13 shows the distribution of the Volta potential across a carbon steel sample scanned under load and at the rest after unloading. It is possible to observe the formation of a low potential (anodic) location in the notch. Unloading passivates the steel surface in the notch, but residual stress in some locations kept them in a relatively active state (Fig. 3.13B). Atomic force microscopy (AFM) revealed that the area with residual plastic stress contains the slip bands of the emergent dislocations (Fig. 3.14A).

The schematic of the strained surface and the formation of the active locations are shown in Fig. 3.14B. It can be supposed that slip bands result in lower potentials. Thus, repassivation results in a less protective oxide film that is altered by dislocations. During the tribocorrosion test, the residual stress can concentrate at the metal surface and decrease the ability of the alloy to re-passivate.

The SKP mapping was carried out for locations on the austenitic stainless steel 301 LN containing residual stress built-in by indentation. In this case, the plastic deformation is also a decreasing surface potential that creates a potential well (Fig. 3.15B). These locations contain residual compressive and tensile shear stresses that also create

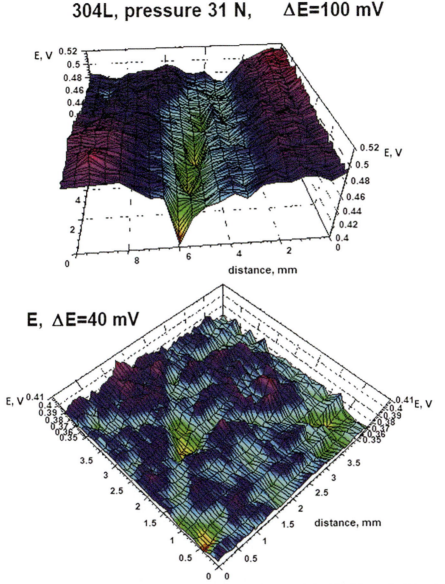

Fig. 3.11 Effect of the scratching on the potential distribution of the 304 SS sample (A) and high strength steel "Hardox" (B). The measurements were performed in air at 60% relative humidity [30]. The scratches are marked by the *lines*.

active surfaces and locations prone to pitting corrosion. Plastic strain creates dislocations and defects at the surface oxide film, which locally decreases the potential of the steel. Thus, locations with compressive residual stress also have low potentials, which can be visualized using SKP [32].

The locations containing pile-ups of dislocations have lower abilities to passivate due to the disordered passive oxide film [31–34]. Due to their electrochemical nature, corrosion reactions are spatially separated and corrosion events follow the distribution of the surface potential. Thus, low potential locations (e.g., the notch in Figs. 3.12 and 3.13 and the potential well in Fig. 3.15) are prone to pitting formation. It is important to note that, during the mechanical stress, the phase transformation of austenite to martensite accumulates internal stress that generates pits at the boundary of the two phases. These new defects in the interfacial regions produce more defective oxides and areas susceptible to pit nucleation [32]. These SKP studies of the effects of stress

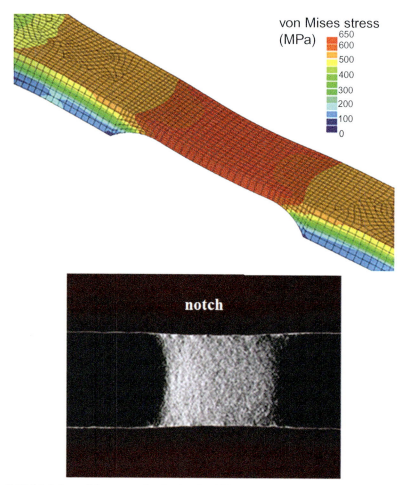

Fig. 3.12 Modeling of the stress distribution for 301 LN stainless steel with notch (A), photograph of the sample with 10% of tensile residual strain (B), potential distribution across steel surface under the load (C) [32, 33].

(Continued)

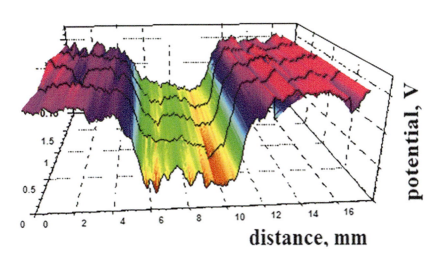

Fig. 3.12, cont'd

on the surface potential are in line with results concerning the effect of cold work on pitting corrosion in steel [35].

The duration of the contact between the active surface and the electrolyte is critical for the stability of the material for stress corrosion cracking and tribocorrosion. Quick repassivation of the wearing surface can decrease the material lost and maintain the integrity of the construction. After grinding (abrasion by emery paper) of the austenitic stainless steel 301LN, it passivates and forms a new oxide film. SKP was applied to the study of the grinded surface (Fig. 3.16). Two parts of the steel surface were grinded and exposed to air for passivation at different times (Fig. 3.16A). These parts obtained different potentials and different levels of passivity. Thus, the passivation of steel in the air is a relatively slow process (Fig. 3.16B). The surface potential is proportional to the thickness of the oxide film [32]. The rates of passivation after different surface treatments can be determined by SKP.

3.2.4 Local electrochemical impedance spectroscopy (LEIS)

Electrochemical impedance spectroscopy (EIS) and Mott-Schottky analysis have been used to analyze the semiconducting properties of passive film on stainless steels (316L and 304) under mechanical stress [36, 37]. The conductivity and capacitance of the oxide film increased when elastic or plastic stress was applied. This effect was explained by an increase in the concentration of donors and acceptors compared to a stress-free state.

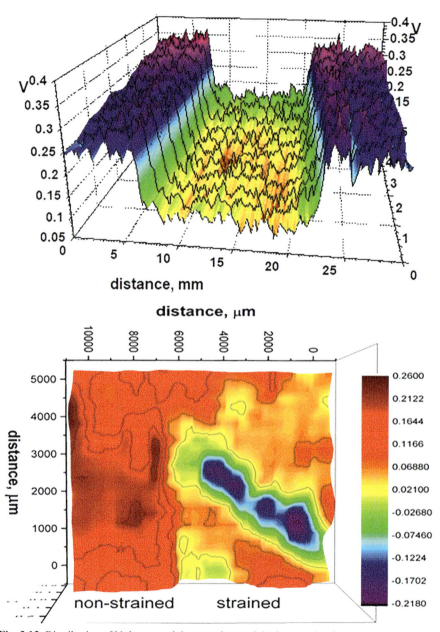

Fig. 3.13 Distribution of Volta potential across the notch in the sample of carbon steel under the tensile stress 310 MPa (A) and at the rest (B) with locations containing residual stress [33].

Mechanical and chemical coupling in tribocorrosion 45

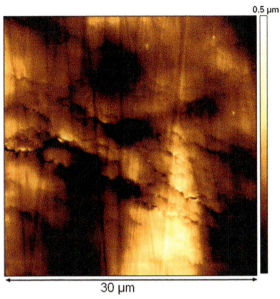

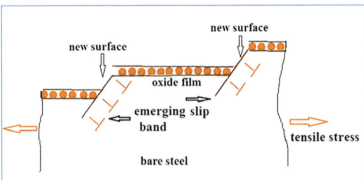

Fig. 3.14 (A) AFM image of the carbon steel surface after application of the tensile stress (Fig. 3.12). The slip steps marked by *arrows*. (B) The schematic of the strained surface [33].

Global EIS gives information on the resistance and capacitance of metal surfaces during tribocorrosion [14–16]. In LEIS, the microelectrode is scanned above the worn surface and supplies information on the oxide film at each location (Fig. 3.17). A comparison of the EIS parameters for the rubbed and bare metals provides direct visualization of the effects of wear. More details on LEIS are given elsewhere [38–40].

LEIS was used to determine the conditions of the passive film after tensile stress application [33]. Fig. 3.18B shows the distribution of the capacitance across the area of the notch and the bare steel of the same sample. The notch contains residual stress and dislocation, slip bands. The capacitance of the mechanically strained area was approximately two to three times higher than that of the nondeformed reference surface. Thus, after steel repassivation, the oxide film has a relatively disordered and defective structure. Under residual stress conditions, the film obtained an increased

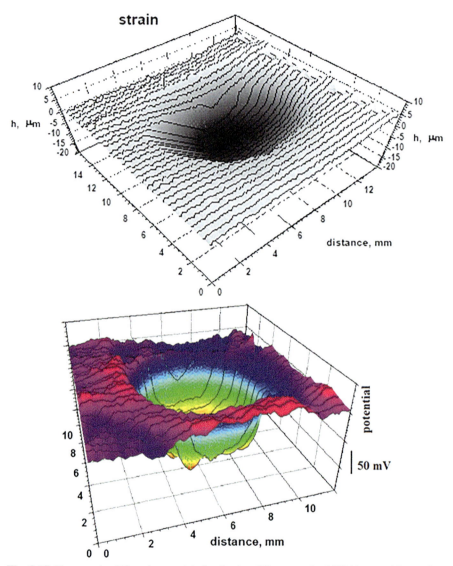

Fig. 3.15 Topography (A) and potential distribution (B) across the 301LN austenitic steel surface with residual stress, after application of local plastic compressive strain 320 MPa (indentation) [32].

concentration of oxygen vacancies that increased its sensitivity to localized attacks, such as pitting corrosion and stress corrosion cracking [33].

3.2.5 Scanning electrochemical microscopy (SECM)

SECM is a powerful technique that helps to correlate microstructural heterogeneities with surface electrochemical reactivity in aqueous electrolytes [41–47]. A disc made of Pt (diameter 10–20 μm) scans the metal surface at the distance of 10–20 μm

Mechanical and chemical coupling in tribocorrosion

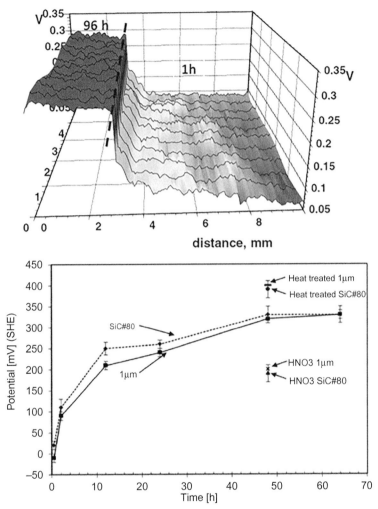

Fig. 3.16 The map of potentials for surface of 301 LN stainless steel after grinding and exposure in the air during 96 and 1 h before measuring (A). Monitoring of the potential 301LN stainless steel in the air after grinding with different roughness and surface treatment (B) [32].

(Fig. 3.19). The tip collects information regarding each location, such as current, potential, and concentration of consumed or released chemical species.

The corrosion of stainless steel, particularly pitting corrosion, is a highly localized process; it can be studied at a micrometer range or less with SECM [43–45]. For pitting corrosion, the potential of the SECM tip is made positive to oxidize the Fe^{2+} ions released from the pit (Fig. 3.19B). Locating the probe above the pit results in an increased current due to the dissolution of the steel. Thus, SECM provides a map of weak locations that are prone to pitting corrosion with high spatial resolution [43].

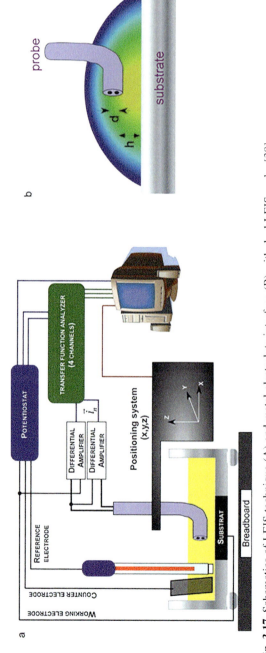

Fig. 3.17 Schematics of LEIS technique (A) and metal-electrolyte interface (B) with dual LEIS probe [38].

Mechanical and chemical coupling in tribocorrosion

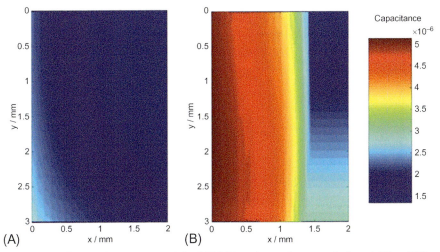

Fig. 3.18 LEIS maps for two parts of the 301 LN steel sample scanned after yielding 10%. (A) Outside the notch and (B) the area across the notch and the bare steel. The measurements were carried out at 0.4 V vs SHE in borate electrolyte [33].

The maps from SECM can be compared with the microstructure or the distribution of intermetallics and inclusions.

SECM was applied to the study of the effect of mechanical stress on the electrochemical reactivity of steel [46–48]. The rate of electron transfer across the metal-electrolyte interface was determined in situ under a load using red-ox mediators. For steel (alloy 800) and thiosulfate-containing electrolyte, it was shown that the formation of the active surface significantly accelerates electrochemical reactions [47].

SECM experiments were performed for several titanium Ti-xCu alloys to characterize the electrochemical reactivity in tribocorrosion experiments [49]. This technique measures a faradic current while the tip is scanned over the surface of the specimen. The microelectrode was a 25-μm diameter platinum wire. It was shown that an increase in the amount of eutectoid at the grain boundaries facilitates the formation of Ti_2Cu intermetallics, thus increasing their hardness. The enlargement of those intermetallics leads to an enhancement of the cathodic reaction rate and a decrease in wear due to their effect on the hardness of the material. During tribocorrosion, galvanic coupling between the worn (active) and unworn (passive) areas was observed. It was pointed out that the reactivity of the surface depends on the stability of the passive conditions [49].

3.2.6 Atomic force microscopy (AFM) and scanning Kelvin probe force microscopy (SKPFM)

AFM is a local technique in which a cantilever scans the surface of a metal and senses its mechanical and electrochemical properties (Fig. 3.20). AFM operates in several modes and can evaluate surface properties, which contribute to tribocorrosion

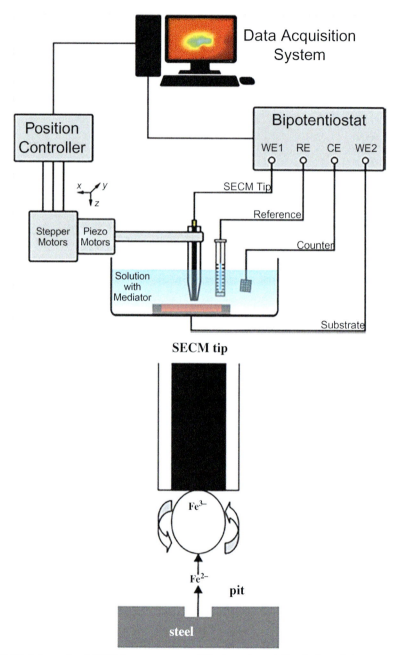

Fig. 3.19 Schematic of SECM measurements (A), detection of dissolution of the steel in the pit (B).

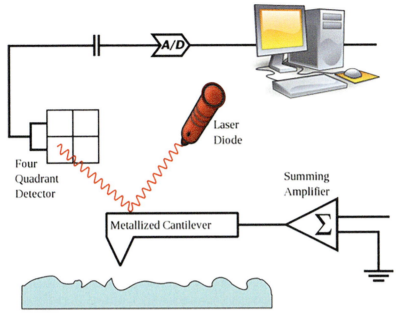

Fig. 3.20 Simplified sketch of AFM. In Kelvin probe force microscopy, a conducting cantilever is scanning over a surface at a constant height in order to map the work function of the surface.

mechanisms at the microlevel and nanolevel (Fig. 3.20). The principle of AFM and examples of its application to tribocorrosion are summarized in Ref. [40].

A nanoindentation technique was developed to evaluate the mechanical properties of surface films with thicknesses in the range of nanometers. Atomic force microscopy (AFM) is a powerful technique because, after testing, it can evaluate the shape of the indent. Nanoindentation is often performed in an electrolyte with potentiostatic electrochemical control. A nanoindentation study revealed that the hardness of the anodic passive film on Ta was higher than the hardness of the metallic substrate [50]. The same study also revealed that the repassivation rate of titanium increases with an increase in the potential difference substrate-solution [50].

For iron crystals, different crystallographic planes have different hardness and different dissolution and passivation rates [40]. As the indentation depth increases during loading, the passive film can be partially ruptured. The passive film, even if very thin, will resist indentation to some extent. Film rupture will promote indentation and film repair will impede indentation. Thus, the nanomechano-electrochemical properties of the passive film are affected by the load-depth curves and can be evaluated [40]. The passivity of stainless steel and the role of chromium ions in steel can be explained at the nanolevel by the high self-healing ability of defective or ruptured locations [40].

A nanoscratching test using AFM in an aqueous electrolyte under potentiostatic control is also useful for evaluating the properties (e.g., the coefficient of friction) of passive metals as a function of crystallographic plains (100) and (110) [50]. High friction coefficients for passive iron and titanium were obtained for in situ scratching relative to ex situ scratching in air.

Kelvin probe force microscopy (SKPFM) is a noncontact variant of AFM (Fig. 3.20). A conducting cantilever is scanned over a surface at a constant height to map the electron work function of the surface. The information evaluated by SKPFM is similar to that evaluated by SKP (see above), but the spatial resolution of the measurements can be close to 100 nm. On the other hand, the calibration procedure and comparison of the data from different measurements are more defined in SKP than in SKPFM. In modern SKP and SKPFM instruments, the topography and distribution of the electron work function can be obtained in one scan simultaneously. Because the work function is influenced by many surface phenomena, including catalytic activity, reconstruction of surfaces, doping, and band bending of semiconductors, charge trapping in dielectrics, and corrosion, the interpretation of results can be difficult. SKPFM is mainly used to determine electrochemical nonuniformity in multiphase alloys that can govern corrosion and electrochemical reactions at the nanolevel.

The stress corrosion cracking mechanism of the AISI 304 stainless steel due to the interaction of the tensile stress with locations of local corrosion (pits) was studied using AFM and SKPFM [51]. The slip deformation, pit growth, and surface potential distribution around the cracks were investigated. The electron backscattering image (SEM), topography (AFM), and map of electron work function (SKPFM) were compared (Fig. 3.21). The decreased electron work function areas were observed on both cracked parts and some the slip deformation parts and disappeared with time when the specimen was kept in the air. These discontinuous cracks joined together and the amount of slip deformation increased with the progress of the crack. It was pointed out that the negative shift of the electron work function corresponds to hydrogen absorption that can lead to local de-cohesion of the alloy. The important role of hydrogen generation in electrochemical reactions and the cracks developing was underlined [51]. This work can be important for tribocorrosion where development of deformation and cracks is often observed (Fig. 3.5).

In the study [52], SKPFM and AFM were applied to analyze the localized corrosion of the aluminum alloy AA2024-T3. An AFM tip was used as a scratching tool to destabilize passivity and accelerate corrosion. SKPFM determined the electrochemical

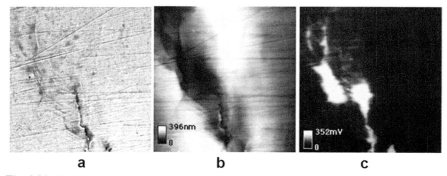

Fig. 3.21 SEM (A), AFM (B), and SKPFM (C) images near the crack tip after19 days of atmospheric corrosion test. Image size is 0.06 mm × 0.06 m [51].

heterogeneity of the surface. The role of intermetallic particles in defined locations on the rates of localized corrosion was determined using SKPFM in the air after exposure to chloride solutions. Al-Cu-Mg particles, which have a noble Volta potential, were actively dissolved in a chloride solution after a certain induction time. Light scratching of the surface by rastering with the AFM tip in contact with the chloride solution resulted in the accelerated dissolution of both pure Al and the alloy 2024-T3 [53]. Thus, abrasion associated with contact AFM in situ resulted in the immediate dissolution of the Al-Cu-Mg particles because of the destabilization of the surface film.

Important for tribocorrosion is the effect of plastic deformation on the corrosion stability of the material at the nanolevel. Duplex 2205-grade stainless steel was studied by SKPFM. In annealed microstructure, the Volta potential difference of 70–90 mV between the austenite and ferrite phases was detected, with ferrite obtaining a net anodic potential. This difference is expected to facilitate microgalvanic coupling between the ferrite and austenite, resulting in the selective dissolution of the ferrite. The introduction of a 40% cold-roll reduction significantly reduced the Volta potential difference between the ferrite and austenite (Fig. 3.22). However, the map of the Volta potentials for the cold-rolled microstructure showed local potential hot spots (highly active sites in confined regions), indicating further corrosion-active sites [54]. This finding is in line with SKP observations of low potential locations containing residual stress [33]. This point agrees with SKP studies of austenite 310LN and carbon steels [32, 33]. The tensile and compressive strains create low potential locations where the residual stress is concentrated (Figs. 3.12–3.14)

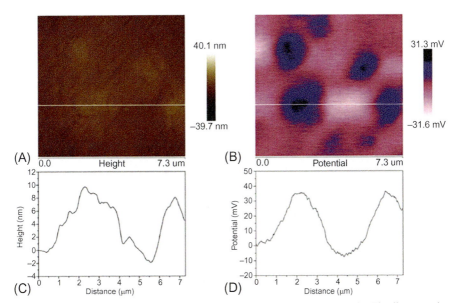

Fig. 3.22 AFM (A) and SKPFM (B) maps for the area of CoCrMo sample. The line scan in topography (C) and Volta potential (D) [53].

Tribocorrosion for medical application is a growing field of study. When a CoCrMo-alloy bioimplant is implanted in a patient, the surface adsorbs biomolecules such as proteins and amino acids. For load-bearing surfaces, some adsorbed proteins can be denatured through tribochemical reactions. The study [55] uses AFM and SKPFM to investigate at the microlevel the local reactions induced by protein adsorption under tribological contact. The results showed that adsorbed molecules could decrease the work function and promote corrosion in CoCrMo alloys. The corrosion products in topography (AFM) correspond to locations in SKPFM map with decreased electron work function (potential) (Fig. 3.23). In the wear track, the albumin denatured and changed the surface potential over time. It was shown that the application of anodic potential the complex protein film can protect the wear track from further damage [55]. Thus SKPFM is a helpful method to study the influence of organics on the tribocorrosion at the microlevel.

3.3 Application of in situ vibrational spectroscopy for studies of tribochemical and tribocorrosion processes

Vibrational spectra can give detailed chemical information on tribofilms and reaction products formed during tribochemical and tribocorrosion processes. The most commonly used techniques Raman and FTIR-spectroscopy provides vibrational spectra of which enable the identification of inorganic and organic molecular species as well as information about the chemical bonding of molecules. Quantification of band intensities makes it possible to follow the generation or consumption of species. Furthermore, detailed analysis of bandshift, bandwidths, and bandshapes can provide additional information about chemical bonding, grain size, and order vs disorder on a bond scale.

Both FTIR- and Raman spectroscopy have used extensively in the field of corrosion for ex situ analysis as well as for in situ studies of corrosion processes and related

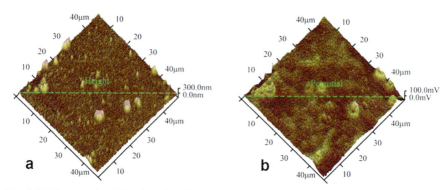

Fig. 3.23 Topography (A) and surface electron work function (potential) image (B) of CoCrMo alloy sample surface after tribocorrosion tests [55].

phenomena [56, 57]. Since the FTIR and Raman spectroscopy are based on the interaction of light with the samples are they relatively easily adapted for in situ studies in electrolyte solutions or the humid air. This is in contrast to methods that use electrons or ions as probing particles, such as photoelectron spectroscopy and secondary ion mass spectroscopy. Vibrational spectroscopy can also be combined with simultaneous electrochemical measurements providing chemical and information about formation or consumption rates of different species as well as rates of electrochemical reactions.

Vibrational spectroscopy is a powerful tool to study buried interfaces such as metal/polymer interfaces or material surfaces in sliding contacts. Experimental set-ups based on infrared attenuated total reflectance (ATR) in combination with electrochemical measurements have been used to study water transport to the metal/polymer interface on coated metal surfaces and subsequent corrosion processes at the interface [58–61].

The buried interfaces of metal/polymer systems are similar to the case with a sliding contact between two bodies in tribological contact in the sense that the interfacial region is not easily accessible for analysis. Olsen et al. [62] used in situ ATR-FTIR to probe tribochemical reactions on diamond-like carbon coatings during rubbing while Piras et al. [63] used FTIR-ATR to study the formation of tribofilms on iron surfaces in the presence of lubricant containing antiwear additives. An experimental set-up for the studies of tribofilm formation has been constructed which makes it possible to record infrared spectra during sliding motion and at the same time measure friction, see Fig. 3.24 [64]. In situ infrared transmission analysis in ball-on flat [65] and pin-on-disc [66] tribometers have also been presented, the latter shown in Fig. 3.25.

Sasaki et al. [67, 68] developed in situ FTIR-ATR microscopical imaging with a focal plane array detector for studies of the time-dependent 2-D distribution of chemical reactions on the friction surfaces, see Fig. 3.26. The time-dependent surface distribution of tricresyl phosphate lubricant at the surface was obtained during sliding motion (Fig. 3.27).

Raman spectroscopy is a powerful technique for studies of tribological processes [69–72] and the technique has several advantages over FTIR-spectroscopy such low-spectra interference from water molecules, more easily adapted for high-temperature studies, and an advantageous wavelength region for studies of oxides, sulfides, and metal-ligand vibrations. An experimental set-up for high temperature in situ Raman tribometry was presented by Muratore et al. [70], as seen in Fig. 3.28. In situ Raman spectroscopy of solid lubricant MoS_2 coating during high temperature (300–700°C) was performed during wear testing. Raman spectroscopy enabled real-time correlation of sliding contact surface chemistry to the measured friction coefficient and showed that the formation of MoO_2 was associated with an increase of the friction coefficient, see Fig. 3.29.

A major challenge with the Infrared and Raman spectroscopy in studying surface processes, such as corrosion and tribocorrosion, is the large-signal contributions from bulk molecules which may overwhelm the signal arising from the species at the surface region. More recently developed techniques such as Vibrational Sum Frequency Spectroscopy (VSFS) and second harmonic generation, based on nonlinear optical effects can probe surface and interfaces without noninterfering contributions from

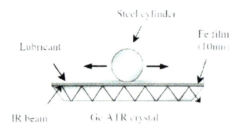

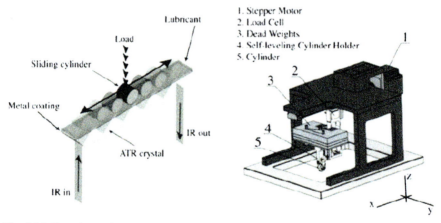

Fig. 3.24 Experimental set-up for in situ ATR-FTIR studies of tribological processes with simultaneous friction measurements [64].

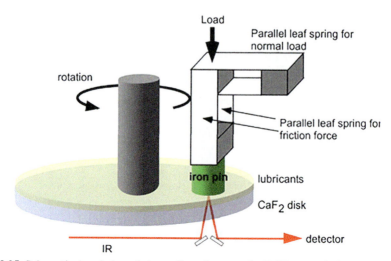

Fig. 3.25 Schematic description of pin-on disc tribometer for FTIR-transmission analysis [66].

Mechanical and chemical coupling in tribocorrosion

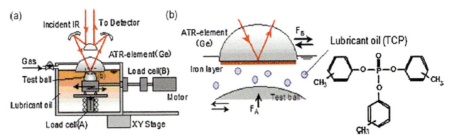

Fig. 3.26 Schematic diagram of FTIR-ATR system for time-dependent FPA imaging of friction surfaces [67, 68].

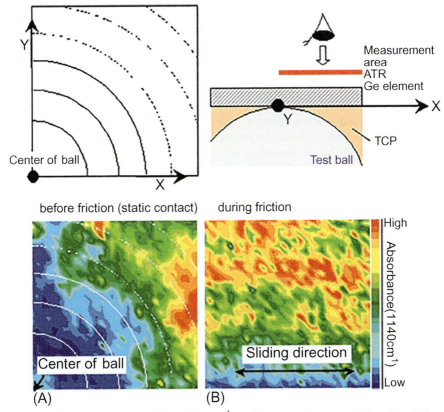

Fig. 3.27 2D-spatial imaging of the 1140 cm^{-1} band intensity for tricresyl phosphate before (A) and during friction (B) [68].

bulk signal contribution [55]. In VSFS are two laser beams temporarily and spatially overlapped on the sample surface. A third beam is then generated with a frequency the same as the frequency of the incoming beams. The generation sum frequency is not allowed in the centrosymmetric media, such as bulk materials. However, at surfaces and interfaces, the symmetry is broken and a Vibrational Sum Frequency signal is

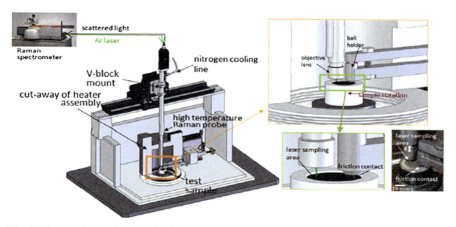

Fig. 3.28 Experimental set-up for in situ Raman spectroscopy for examination of high-temperature tribological processes [70].

generated. This makes the VSFS an inherently specific technique. The VSFS technique has been used to study corrosion phenomena under in situ atmospheric conditions [56, 73] but also for investigations of processes related to tribology [66, 74].

Microscopy and chemical imaging vibrational spectroscopy techniques based on optical microscopes connected to FTIR and Raman spectrometers make them powerful techniques for analysis and chemical imaging on a micrometric level. However, the lateral resolution of these techniques is diffraction-limited which means that the length scale below the micrometer range is not accessible for microscopy and chemical imaging by the conventional FTIR and Raman microscopy. However, more recent developments in the scanning probe spectroscopical techniques, tip-enhanced Raman spectroscopy (TERS) and nanoscale infrared spectroscopy (Nano-IR) such as scattering scanning near field optical microscopy (sSNOM) has extended the lateral resolution to subnanometer region and the order of 10 nm, respectively [75, 76]. In these techniques, the electric field strength of the incoming light is enhanced at the metal-coated metal AFM tip. Thus, the light field is localized at the nanoscale tip and the object and the spatial resolution is dependent on the tip curvature and not on the optical diffraction limit in an objective. Fig. 3.30 shows an illustration of the principle of the nano-IR technique with top-down illumination. sSNOM was employed for studies of the distribution of corrosion products on copper exposed to formic acid by Johnson and Böhmler [77] and it was shown that the technique can provide spectral information of a corroded metal surface with a spatial resolution of 20 nm. Infrared microscopy at the nanoscale can also be obtained by other techniques, such as photothermal-induced resonance (PTIR) which is based on local detection of the thermal expansion of the sample surface with the AFM tip, when a pulsed tunable laser source is focused on a sample [78].

Mechanical and chemical coupling in tribocorrosion

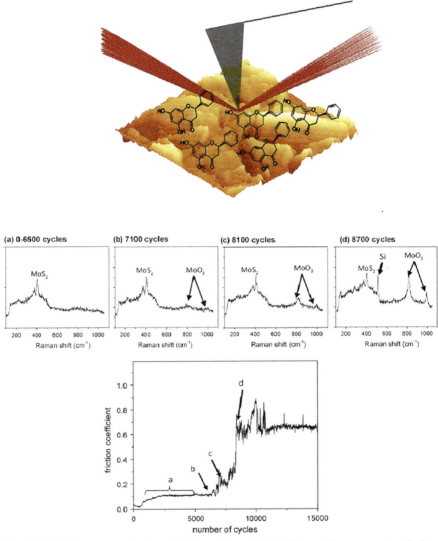

Fig. 3.29 Friction test results for a MoS2 coating at 330–350°C and Raman spectra obtained after different number of sliding cycles [70].

3.4 Conclusions

In this paper, local electrochemical methods were discussed in connection to tribocorrosion. The main objective of these techniques is to observe at the microscale the surface distribution of corrosion-related parameters during and after tribological action. To understand the tribocorrosion mechanism, local electrochemical data have

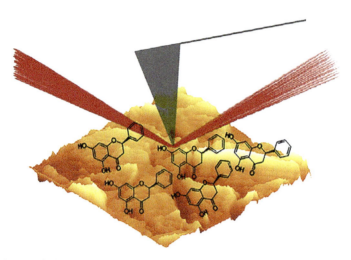

Fig. 3.30 Schematic illustration of nano-IR spectroscopy with top-down illumination [76].

to be correlated with the surface phase composition, microstructure, grains and inclusions, intermetallics, stress distribution, and concentrators. It was demonstrated that the local modern electrochemical techniques SRET-SVET, LEIS, scanning microcell, SECM, SKP, AFM, and SKPFM can supply different kinds of spatially resolved information related to the corrosion and wearing of surfaces.

SRET-SVET is useful for obtaining the distribution of the corrosion currents flowing a few hundred micrometers above the corroding surface in an electrolyte. The anodic and cathodic locations can be determined and compared with the distribution of the surface phases. The kinetic passivation at different locations after tribological contact can be determined.

LEIS gives information about the oxide film in the form of distribution the capacitance and surface film resistance. The impact of tribocorrosion in an aqueous electrolyte on depassivation and repassivation can be determined as a function of dislocations and residual stress distributions.

The scanning microcell technique makes it possible to obtain the distribution of electrochemical parameters across the entire surface. The presence of an abrading partner in the cell makes it possible to determine the rates of activation-passivation at each location (e.g., the grain) during tribocorrosion.

SECM studies the surface reactivity, such as the rate of electron transfer across the metal-electrolyte interface. The efficiency of cathodic and anodic reactions can be evaluated at different locations and for different mechanical actions (e.g., straining).

SKP analyzes, at the microlevel, give the electrochemical heterogeneity (Volta potential) of the surface in the air or another gaseous atmosphere. It can effectively determine the effect of tribocorrosion under a thin electrolyte film in atmospheric conditions. The distribution of the electrochemical potential and the electron work function show the locations with decreased passivity due to the presence of dislocations, high concentrations of residual stress, or other surface defects and heterogeneities.

AFM-SKPFM analyzes the properties of the corroding surface at the nanolevel. The topography and distribution of the Volta potential give the anodic and cathodic locations as functions of the alloy surface composition, microstructure, and mechanical work such as straining or rubbing.

Vibrational spectroscopy is a powerful tool to obtain chemical information about corrosion processes under various in situ conditions. Experimental set-ups for in situ tribometry based on FTIR and Raman spectroscopy have been designed to obtain information about the formation of tribofilms and reaction products with simultaneous friction measurements. Although most in situ FTIR and Raman studies are performed for studies of lubrication and very few in the field of tribocorrosion, the techniques are potentially also very interesting for studies of tribocorrosion phenomena, in particular in combination with electrochemical measurements.

New developments, such as tip-enhanced Raman Spectroscopy and nano-IR have made it possible to make microscopical measurements with a lateral resolution in the nanometric scale. These techniques could be very useful for the correlation of chemical composition of reaction products and tribofilms to AFM-SKPFM studies of localized electrochemical processes and AFM and electron microscopy studies of tribocorrosion attacks.

3.5 List of symbols

A	albumin concentration in aqueous solution (g/L)
d	distance (m)
F	Faraday's constant, 96,485 C/mol
I	corrosion current
K_c	total weight loss due to corrosion degradation process (g or mm^3)
K_{co}	weight loss in absence of mechanics, pure dissolution without any mechanical degradation
K_w	total weight loss due to mechanical degradation (g or mm^3)
K_{wc}	total weight loss (g or mm^3)
K_{wo}	weight loss in absence of corrosion, pure friction without solution
M	molar mass (g/mol)
N	number of exchanged electrons
R_1	$= \frac{K_c}{K_w}$
R_2	$= \frac{\Delta K_w}{\Delta K_c}$
W	total wear volume (mm^3)
W_c	pure corrosive wear (mm^3)
W_m	pure mechanical wear (mm^3)
W_{tr}	wear volume related to the wear track area (10^{-5} mm^3)
δI or ΔI	current drop due to friction/fretting (µA, most of the time)
ΔK_c	effect of wear on the weight loss due to corrosion
ΔK_w	effect of corrosion on the weight loss due to wear
Δt	duration of measuring of the current (s)
$\Delta W_{cm} + \Delta W_{syn}$	corrosive part on mechanical wear and the synergistic term (mm^3)

ΔW_{cm} corrosive part on mechanical wear (mm^3)
ΔW_{mc} mechanical part on corrosive wear (mm^3)

3.6 List of abbreviations

PEO plasma electric oxidation
R4, R5 name of various steel grade
TSA thermally sprayed aluminum

References

[1] D. Landolt, S. Mischler (Eds.), Tribocorrosion of Passive Metals and Coatings, Woodhead Publishing Series in Metals and Surface Engineering 2011, 2011.
[2] S. Mischler, A.I. Munoz, Wear of CoCrMo alloys used in metal-on-metal hip joints: a tribocorrosion appraisal, Wear 297 (2013) 1081–1094.
[3] J.F. Archard, Contact and rubbing of flat surfaces, J. Appl. Phys. 24 (1953) 981.
[4] V. Guruswamy, J.O.'.M. Bockris, Triboelectrochemistry, in: J.O.'.M. Bockris, B.E. Conway, E. Yeager, R.E. White (Eds.), Compressive Treatise of Electrochemistry, vol. 4, Plenum Press, New York, London, 1981, p. 463.
[5] M.N.F. Ismail, T.J. Harvey, J.A. Wharton, R.J.K. Wood, A. Humphreys, Surface potential effects on friction and abrasion of sliding contacts lubricated by aqueous solutions, Wear 267 (2009) 1978–1986.
[6] F.J. Hen, Y.Z. Fang, S.J. Jin, The study of corrosion–wear mechanism of Ni–W–P alloy, Wear 311 (2014) 14–20.
[7] Y. Sun, V. Rana, Tribocorrosion behaviour of AISI 304 stainless steel in 0.5 M NaCl solution, Mater. Chem. Phys. 129 (2011) 138–147.
[8] P. Ponthiaux, F. Wenger, D. Drees, J.-P. Celis, Electrochemical techniques for studying tribocorrosion processes, Wear 256 (2004) 459–468.
[9] S. Mischler, Triboelectrochemical techniques and interpretation methods in tribocorrosion: a comparative evaluation, Tribol. Int. 41 (2008) 573–583.
[10] S. Mischler, A. Spiegel, M. Stemp, D. Landolt, Influence of passivity on the tribocorrosion of carbon steel in aqueous solutions, Wear 251 (2001) 1295–1307.
[11] N. Espallargas, R. Johnsen, C. Torres, A.I. Muñoz, Galvanic interaction of the worn and unworn surfaces are studying using zero ammeter, Wear 307 (2013) 190–197.
[12] Y. Zhang, X.-Y. Yin, F.-Y. Yan, Tribocorrosion behaviour of type S31254 steel in seawater: identification of corrosion wear components and effect of potential, Mater. Chem. Phys. 179 (2016) 273–281.
[13] A. Dalmau, C. Richard, A. Igual Munoz, Degradation mechanisms in martensitic stainless steels: wear, corrosion and tribocorrosion appraisal, Tribol. Int. 121 (2018) 167–179.
[14] M. Keddam, P. Ponthiaux, V. Vivier, Tribo-electrochemical impedance: a new technique for mechanistic study in tribocorrosion, Electrochim. Acta 124 (2014) 3–8.
[15] J. Geringer, J. Pellier, M.L. Taylor, D.D. Macdonald, Electrochemical impedance spectroscopy: insights for fretting corrosion experiments, Tribol. Int. 68 (2013) 67–76.
[16] J. Geringer, B. Normand, C. Alemany-Dumont, R. Diemiaszonek, Assess in the tribocorrosion behaviour of Cu and Al by electrochemical impedance spectroscopy, Tribol. Int. 43 (2010) 1991–1999.

[17] M. Keddam, F. Wenge, Electrochemical methods in tribocorrosion, in: Tribocorrosion of Passive Metals and Coatings, Woodhead Publishing Series in Metals and Surface Engineering, Woodhead Publishing, 2011, pp. 187–221.
[18] Y.N. Kok, R. Akid, P.E. Hovsepian, Tribocorrosion testing of stainless steel (SS) and PVD coated SS, Wear 259 (2005) 1472–1481.
[19] I. Annergren, F. Zou, D. Thierry, Application of localized electrochemical techniques to study kinetics of initiation and propagation during pit growth, Electrochim. Acta 44 (1999) 4383–4389.
[20] B. Vuillemin, X. Philippe, R. Oltra, V. Vignal, L. Coudreuse, L.C. Dufour, E. Finot, SVET, AFM and AES study of pitting corrosion initiated on MnS inclusions by microinjection, Corros. Sci. 45 (2003) 1143–1159.
[21] R. Akid, P. Roffey, D. Greenfield, D. Guillen, Application of scanning vibrating electrode technique (SVET) and scanning droplet cell (SDC) techniques to the study of weld corrosion, in: Local Probe Techniques for Corrosion Research, CRC Press, 2013, ISBN: 978-1-4200-5405-7.
[22] T. Prosek, A. Nazarov, H.B. Xue, S. Lamaka, D. Thierry, Role of steel and zinc coating thickness in cut edge corrosion of coil-coated materials in atmospheric weathering conditions; part 1: laboratory study, Prog. Org. Coat. 99 (2016) 356–366.
[23] Z.Y. Liu, X.G. Li, Y.F. Cheng, In-situ characterization of the electrochemistry of grain and grain boundary of an X70 steel in a near-neutral pH solution, Electrochem. Commun. 12 (2010) 936–953.
[24] A.B. Oliveira, A.C. Bastos, C.M. Fernandes, C.M.S. Pinho, A.M.R. Senos, E. Soares, J. Sacramento, M.L. Zheludkevich, M.G.S. Ferreira, Corrosion behaviour of WC-10% AISI 304 cemented carbides, Corros. Sci. 100 (2015) 322–331.
[25] F. Assi, H. Bohni, Study of wear-corrosion synergy with a new microelectrochemical technique, Wear 233–235 (1999) 505–514.
[26] H. Bohni, T. Suter, F. Assi, Micro-electrochemical techniques for studies of localized processes on metal surfaces in the nanometer range, Surf. Coat. Technol. 130 (2000) 80–86.
[27] V. Vignal, N. Mary, P. Ponthiaux, F. Wenger, Influence of friction on the local mechanical and electrochemical behaviour of duplex stainless steels, Wear 261 (2006) 947–953.
[28] Y. Takabatake, K. Fushimi, T. Nakanishi, Y. Hasegawa, Grain dependent passivation of iron in sulfuric acid solution, J. Electrochem. Soc. 161 (2014) C594–C600.
[29] M. Stratmann, H. Streckel, On the atmospheric corrosion of metals, which are covered with thin electrolyte layers. 1. Verification of electrochemical technique, Corros. Sci. 30 (1990) 681–689.
[30] A. Nazarov, D. Thierry, Application of Volta potential mapping to determine metal surface defects, Electrochim. Acta 52 (2007) 7689–7696.
[31] G. Bäck, A. Nazarov, D. Thierry, Localized corrosion of heat—treated and welded stainless steel studied by scanning Kelvin probe, Corrosion NACA 61 (2005) 951–960.
[32] N. Fuertes Casals, A. Nazarov, F. Vucko, R. Pettersson, D. Thierry, Influence of mechanical stress on the potential distribution on a 301 LN stainless steel surface, J. Electrochem. Soc. 162 (2015) C465–C472.
[33] A. Nazarov, V. Vivier, D. Thierry, F. Vucko, B. Tribollet, Effect of mechanical stress on the properties of steel surfaces: scanning kelvin probe and local electrochemical impedance study, J. Electrochem. Soc. 164 (2017) C66–C74.
[34] A. Nazarov, F. Vucko, D. Thierry, Scanning kelvin probe for detection of the hydrogen induced by atmospheric corrosion of ultra-high strength steel, Electrochim. Acta 216 (2016) 130–139.

[35] L. Peguet, B. Malki, B. Baroux, Influence of cold working on the pitting corrosion resistance of stainless steels, Corros. Sci. 49 (2007) 1933–1948.
[36] V. Vignal, C. Valot, R. Oltra, M. Verneau, L. Coudreuse, Analogy between the effects of a mechanical and chemical perturbation on the conductivity of passive films, Corros. Sci. 44 (2002) 1477–1496.
[37] C.M. Rangel, T.M. Silva, M. da Cunha Belo, Semiconductor electrochemistry approach to passivity and stress corrosion cracking susceptibility of stainless steels, Electrochim. Acta 50 (2005) 5076–5082.
[38] V.M.-W. Huang, S.-L. Wu, M.E. Orazem, N. Pébère, B. Tribollet, V. Vivier, Local electrochemical impedance spectroscopy: a review and some recent developments, Electrochim. Acta 56 (2011) 8048–8057.
[39] F. Zou, D. Thierry, H.S. Isaacs, A high-resolution probe for localized electrochemical impedance spectroscopy measurements, J. Electrochem. Soc. 144 (1997) 1957–1965.
[40] P. Marcus, F. Mansfeld (Eds.), Analytical Methods in Corrosion Science and Engineering, CRS Taylor and Francis, Boca Raton, London, New York, Singapore, 2006.
[41] R.G. Kelly, J.R. Scully, D.W. Shoesmith, R.G. Buchheit, Electrochemical Techniques in Corrosion Science and Engineering, Marcel Dekker, New York, 2003, pp. 285–288.
[42] A.J. Bard, F.R.F. Fan, D.T. Pierce, P.R. Unwin, D.O. Wipf, F.M. Zhou, Chemical imaging of surfaces with the scanning electrochemical microscope, Science 25 (1991) 468.
[43] Y.Y. Zhu, D.E. Williams, Scanning electrochemical microscopic observation of a precursor state to pitting corrosion of stainless steel, J. Electrochem. Soc. 144 (1997) L43.
[44] D.O. Wipf, Initiation and study of localized corrosion by scanning electrochemical microscopy, Colloids Surf. A Physicochem. Eng. Asp. 93 (1994) 251.
[45] Y.H. Yin, L. Niu, M. Lu, W.K. Guo, S.H. Chen, In situ characterization of localized corrosion of stainless steel by scanning electrochemical microscope, Appl. Surf. Sci. 255 (2009) 9193.
[46] R.K. Zhu, J.L. Luo, Investigation of stress-enhanced surface reactivity on Alloy 800 using scanning electrochemical microscopy, Electrochem. Commun. 12 (2010) 1752.
[47] D. Sidane, M. Touzet, O. Devos, M. Puiggali, J.P. Larivière, J. Guitard, Investigation of the surface reactivity on a 304L tensile notched specimen using scanning electrochemical microscopy, Corros. Sci. 87 (2014) 312.
[48] P. Sun, Z. Liu, H. Yu, M.V. Mirkin, Effect of mechanical stress on the kinetics of heterogeneous electron transfer, Langmuir 24 (2008) 9941–9944.
[49] V. Guinón Pina, V. Amigó, A. Igual Munoz, Microstructural, electrochemical and tribo-electrochemical characterisation of titanium-copper biomedical alloys, Corros. Sci. 109 (2016) 115–125.
[50] M. Seo, Y. Kurata, Nano-mechano-electrochemical properties of passive titanium surfaces evaluated by in-situ nano-indentation and nano-scratching, Electrochim. Acta 49 (2003) 3221–3228.
[51] H. Masuda, SKFM observation of SCC on SUS304 stainless steel, Corros. Sci. 49 (2007) 120–129.
[52] P. Schmutz, G.S. Frankel, Corrosion study of AA2024-T3 by scanning Kelvin probe force microscopy and in situ atomic force microscopy scratching, J. Electrochem. Soc. 145 (1998) 2295–2306.
[53] Z. Wang, Y. Yan, L. Xing, Y. Su, L. Qiao, The role of hard phase carbides in tribocorrosion processes for a Co-based biomedical alloy, Tribol. Int. 113 (2017) 370–376.
[54] C. Örnek, D.L. Engelberg, SKPFM measured Volta potential correlated with strain localisation in microstructure to understand corrosion susceptibility of cold-rolled grade 2205 duplex stainless steel, Corros. Sci. 99 (2015) 164–171.

[55] Y. Yu, H. Yang, Y. Su, L. Qiao, Study of the tribocorrosion behaviors of albumin on a cobalt-based alloy using scanning Kelvin probe force microscopy and atomic force microscopy, Electrochem. Commun. 64 (2016) 61–64.
[56] S. Hosseinpour, M. Johnson, Vibrational spectroscopy in studies of atmospheric corrosion, Materials 10 (2017) 413.
[57] A. González-Orive, I. Giner, T. de los Arcos, A. Keller, G. Grundmeier, Analysis of polymer/oxide interfaces under ambient conditions—an experimental perspective, Appl. Surf. Sci. 442 (2018) 581–594.
[58] K. Wapner, M. Stratmann, G. Grundmeier, In situ infrared spectroscopic and scanning Kelvin probe measurements of water and ion transport at polymer/metal interfaces, Electrochim. Acta 51 (2006) 3303–3315.
[59] M. Öhman, D. Persson, C. Leygraf, In situ ATR-FTIR studies of the aluminium/polymer interface upon exposure to water and electrolyte, Prog. Org. Coat. 57 (2006) 78–88.
[60] M. Öhman, D. Persson, An integrated in situ ATR-FTIR and EIS set-up to study buried metal–polymer interfaces exposed to an electrolyte solution, Electrochim. Acta 52 (2007) 5159–5171.
[61] P. Taheri, J.H.W. de Wit, H. Terryn, J.M.C. Mol, In situ study of buried metal–polymer interfaces exposed to an aqueous solution by an integrated ATR-FTIR and electrochemical impedance spectroscopy system, J. Phys. Chem. C 117 (2013) 20826–20832.
[62] J.E. Olsen, T.E. Fischer, B. Gallois, In situ analysis of the tribochemical reactions of diamond-like carbon by internal reflection spectroscopy, Wear 200 (1996) 233–237.
[63] F.M. Piras, A. Rossi, N.D. Spencer, Combined in situ (ATR FT-IR) and ex situ (XPS) study of the ZnDTP-iron surface interaction, Tribol. Lett. 15 (2013) 181–191.
[64] F. Mangolini, A. Rossi, N.D. Spencer, In situ attenuated total reflection (ATR/FT-IR) tribometry: a powerful tool for investigating tribochemistry at the lubricant–substrate interface, Tribol. Lett. 45 (2012) 207–218.
[65] M. Cann, H.A. Spikes, In-contact IR spectroscopy of hydrocarbon lubricants, Tribol. Lett. 19 (2005) 289–297.
[66] K. Miyake, T. Kume, M. Nakano, A. Korenaga, K. Takiwatari, R. Tsuiboi, S. Sasaki, Effects of surface chemical properties of self-assembled monolayers lubricated with oleic acid, Tribol. Online 7 (2012) 218–224.
[67] K. Sasaki, N. Inayoshi, K. Tashiro, Development of new in-situ observation system for dynamic study of lubricant molecules on metal friction surfaces by 2D fast-imaging FTIR-ATR spectrometer, Rev. Sci. Instrum. 79 (2008) 123702.
[68] K. Sasaki, N. Inayoshi, K. Tashiro, Friction-induced dynamic chemical changes of tricresyl phosphate as lubricant additive observed under boundary lubrication with 2D fast imaging FTIR-ATR spectrometer, Wear 268 (2010) 911–916.
[69] T.W. Scharf, I.L. Singer, Monitoring transfer films and friction instabilities with in situ Raman tribometry, Tribol. Lett. 14 (2003) 3–8.
[70] C. Muratore, J.E. Bultman, S.M. Aouadi, A.A. Voevodin, In situ Raman spectroscopy for examination of high temperature tribological processes, Wear 270 (2011) 140–145.
[71] J. Cheng, T. Wang, Z. Chai, X. Lu, Tribocorrosion study of copper during chemical mechanical polishing in potassium periodate-based slurry, Tribol. Lett. 58 (2015) 8.
[72] M. Miayima, K. Kitamura, K. Matsumoto, In situ Raman tribometry for the formation and removal behavior of FeS_2 tribofilm in the scuffing process, Tribol. Online 11 (2) (2016) 268–388.
[73] J. Hedberg, J. Henriquez, S. Baldelli, C.M. Johnson, C. Leygraf, Initial atmospheric corrosion of zinc exposed to formic acid, investigated by in situ sum frequency spectroscopy and density functional theory, J. Phys. Chem. C 113 (2008) 2088–2095.

[74] A. Ghalgaoui, R. Shimizu, S. Hosseinpour, R. Álvarez-Asencio, C.T. McKee, C. Johnson, M. Rutland, Monolayer study by VSFS: in situ response to compression and shear in a contact, Langmuir 30 (2014) 3075–3085.
[75] N. Jiang, D. Kurouski, E.A. Pozzi, N. Chiang, M.C. Hersam, R.P. Van Duyne, Tip-enhanced Raman spectroscopy: from concepts to practical applications, Chem. Phys. Lett. 659 (2016) 16–24.
[76] M. Denisov, A. Jenistov, A. Parchanská-Kokaislová, P. Matejka, V. Prokopec, M. Svecová, The use of infrared spectroscopic techniques to characterize nanomaterials and nanostructures: a review, Anal. Chim. Acta 1031 (2018) 1–14.
[77] C.M. Johnson, M. Böhmler, Nano-FTIR microscopy and spectroscopy studies of atmospheric corrosion with a spatial resolution of 20 nm, Corros. Sci. 108 (2016) 60–65.
[78] A. Dazzi, C.B. Prater, AFM-IR: technology and applications in nanoscale infrared spectroscopy and chemical imaging, Chem. Rev. 117 (2017) 5146–5173.

Managing tribocorrosion investigations by stress mapping: Dual mobility concept, hip implant, as required step

4

Jean Geringer[a] and Caroline Richard[b]
[a]Université de Lyon, IMT Mines Saint-Etienne, Centre CIS, INSERM SainBioSE U1059, Saint-Etienne, France, [b]Université de Tours, GREMAN UMR CNRS 7347/Polytech Tours, Tours, France

4.1 Introduction

The previous chapter is dealing with the ex situ technique to investigate tribocorrosion. Knowing some information from mechanics is the first task to do to investigate some tribocorrosion issues. The contact mechanics have this role. Every contact would be not regular: not sphere-plane, or cylindrical-plane, etc. About conformal contact, sphere/sphere as an example, Hertz's theory gives some limitations and a key one is the influence of roughness. Modeling/analytical calculations reached their limits; let's go back to the beginning by measuring the contact pressure in live conditions. Some devices were elaborated (FujiFilm prescale) and some are ongoing (Tekscan) stresses mapping technology. Before describing the tool, one should make sense by describing the case study: the dual mobility total hip prosthesis.

A hip implant has been patented for the first time in 1976 by Pr. Gilles Bousquet (Saint-Etienne/France). As usual, during 20 years, the patent was deposited thanks to a company. Thus, the other ones (hip implants competitors) did denigrate the idea and the concept at the same time. Twenty years later, the same competitors manufactured accurately the same implant, i.e., dual mobility hip implant, and they found the concept like awesome... The concept may be summarized into two mottos: little head diameter (low friction/less debris) and large mobility (head/cup, 1st mobility, and cup/metal back, dual mobility-2nd mobility), Fig. 4.1A and B.

At the beginning of the story, it was an issue of friction: two interfaces mean more wear, more debris, more toxicity, it makes sense... However, Bousquet's idea was: "when the first mobility is out, contact between the neck and the insert; the second mobility is triggering and the movement, mobility, is going further." The clinical outcomes were beneficial from the dual mobility concept. It was in 2011, work from explants, the comparison between single and dual was practiced and no difference concerning the wear volume. The pristine Bousquet's idea was right. The cup was

Mechanical and Electro-chemical Interactions Under Tribocorrosion. https://doi.org/10.1016/B978-0-12-823765-6.00004-3
Copyright © 2021 European Federation of Corrosion. Published by Elsevier Ltd. All rights reserved.

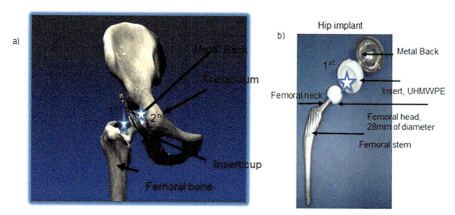

Fig. 4.1 (A) Scheme about a hip prosthesis, dual mobility, 1st and 2nd; (B) Hip implant prosthesis, single mobility 1st.

made of ultrahigh molecular weight polyethylene (UHMWPE). Tribocorrosion is a topic of interest because of friction between metal and polymer.

The first concern of tribocorrosion issue is determining the stresses involved in the contact between two materials. Due to the polyethylene roughness, using some modeling to predict the contact stresses involves a multiscale issue: roughness is a question of micrometers and the size is some centimeter scale. Moreover, the contact has to be considered in solution (bovine serum through in vitro tests). Knowing the liquid thickness, locally, is submitted to the roughness impact. Practicing the experiment sounds as a requirement, in an actual case. Thus to have some information on live stress measurement, the methodology will be emphasized through the first step: choosing the right pressure transducer.

The first experimental tool to measure localized stresses developed in 1996 by Hasler et al. [1] was the pressure-sensitive film approach.

The two sheets are usually composed of polyurethane composite. The transfer sheet has microcapsule bubbles (2- to 26-μm diameter) which encapsulates a colorless liquid. The developer sheet contains a medium reacting with the liquid released from the transfer sheet resulting in a red or pink stain.

That technology is based on ink capsules that are affected by pressure, e.g., they turn red under 200–300 MPa and green under 20–40 MPa. Fig. 4.2.a shows a schematic structure of Fujifilm (FF-Prescale, www.fujifilm.com): when the sheets are pressed together between two articulating objects, the transfer sheet's microcapsules burst, releasing a mild acid, which reacts with the developer sheet's indicator layer. The resulting red-colored contact area has a size, shape, and density, pressure distribution, Fig. 4.2B [2, 3], that depends on applied force, loading rate, loading duration, interfacial geometry, temperature, and humidity. High-resolution digital image analysis techniques can be used for recovering these quantitative data.

Sensitive pressure FF-prescale is the most popular tool for measurement of contact area and stress in articular joints since the end of the 1990's. They were frequently

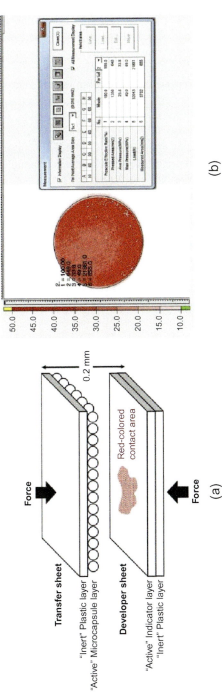

Fig. 4.2 Structure of FF (A) sandwich structure; (B) Example of FF-prescale imaging software window showing a circular contact area. The tabulated parameters include total contact area, average contact stress, maximum contact stress, and applied force [2].

used in several works for hip and knee joints because they are efficient, convenient. However, the authors highlighted the fact that the calibration step has a moderate accuracy in the case of curved surfaces, like hip joints as an example [4, 5]. Indeed, the FF-prescale can crinkle when used for highly curved geometries, and it is sometimes difficult to accurately detect pressure gradients near the edges of the contact area. Pressure-sensitive FF-prescale can be used to record joint stresses under static loading conditions, and there are also methods to measure cartilage-on-cartilage contact through activities of daily living (pivoting, stair climbing, walking, etc.), using various spectroscopies, as X-ray. These methods provide some insight into understanding joint contact mechanics, they are only suitable for quasistatic loading. This method is limited to in vitro laboratory usage. It requires minimum pressure threshold to detect contact. Its spatial resolution is restricted. It can be sensitive to shear. Its finite thickness may interfere somewhat with actual interfacial contact mechanics.

To mitigate these various misfits, another method is available as Tekscan sensors (www.tekscan.com). They allow direct measurement of interfacial contact area and stress, and they are already being used for a variety of robotic, automotive, ergonomic, and biomedical applications. In particular, Tekscan sensors (i.e., K-scan) are useful to measure changes in joint contact mechanics in healthy, pathologic, and artificial joints under quasistatic and dynamic loading conditions. The pressure mapping is based on the piezoresistance technology. The Tekscan sensors consist of a matrix of row and column electrodes created by printing conductive silver ink onto a thin polyester film substrate. The electrodes are further coated with a piezoresistive ink such that a sensing cell is created at each intersection point in the matrix, Fig. 4.2. Changes in the resistance of the cells are measured through scanning electronics that connect to the tail of the sensor. The sensors are thin (100- to 200-μm thick). As the result, this resistive technology helps reduce changes in joint kinematics. These components can measure high stress (up to 60 MPa/cell) where average stress in the joint can reach up to 8–11 MPa/cell.

Under pressure (applied stresses), the dimensions of a wire (length and diameter) can vary, so that the monitored resistance is changing according to the following physics relation:

$$R = \frac{\rho \cdot l}{S} \tag{4.1}$$

where R is the resistance (Ω); ρ is the resistivity (Ω m^2/m); L is the length (m); and S is the area (m^2). On applying a controlled voltage on each wire with an ammeter, the intensity is measured. At that time, one stress value is corresponding to a measured intensity.

Finally, a calibration step was managed with the device and the experimental tool. Thus, this step was managed in the case of any transducer. Some calibrated dead weights were applied on the transducer and some monitored loads were adjusted through it. It is worth noting, at that time, that the procedure does contain some statistics treatment to improve the accuracy of measurements, from minimum to maximum measured values, during this calibration step. Every experimental data would suffer from this critical stage. Every transducer needs usually this required step. Nothing new on the wall was described from there, that stage is related to the experiments managing....

Under applied loads, the pressure-sensitive film inserted between spheres in contact is exposed to shearing, folding, and bending. The pressure-sensitive film consists of wires embedded in a polymer matrix that allows maintain it, Fig. 4.3. However, due to some mechanical solicitations, the adhesion between wires and polymer matrix can be perturbed. One may pay attention to using resistive technology to account for issues associated with the creep of the resistive ink, sensor shear, and wrinkling of the sensor due to the inherent stiffness of the polyester sheets. They are also prone to drift, and the results are not reproducible without drastic experimental care. Applying methodology *ad hoc* is required to take into account these inherent inaccuracies.

This set-up was investigated through this study, Fig. 4.4. It allows to measure in situ stresses in real configurations and during dynamic rotation, i.e., during abduction/adduction and flexion/extension phases. It allows also to measure stresses under friction conditions and in presence of liquids in-between the mapping film/head and the cup/mapping film interfaces. To protect the pressure-sensitive film from liquid

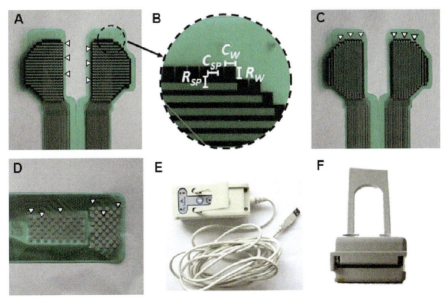

Fig. 4.3 Different Tekscan components. (A) Top of the sensor. *White arrows* show the location of the connection between the rows of the sensing elements and the electrical leads to the handle connector, (B) a zoomed-in view of the top of the sensor showing measurements of the column width CW, row width RW, column spacing CSP, and row spacing RSP, (C) bottom of the sensor with some of the column sensing element connections indicated with *white arrows*, (D) the handle connector end of the sensor with *white arrows* indicating the location of some of the row and column connections for the Tekscan handle, (E) Tekscan handle top view, and (F) Tekscan handle side view [3].

Fig. 4.4 Tekscan K-scan 4402 pressure sensitive film.

adsorption/absorption, a protective film was inserted to cover the mapping film. This protective film is made of polyvinylidene chloride (a thin film of 0.5 mm, wrapping film in food packaging). On adding some other interfaces, the accuracy and the repeatability need to be cautiously considered [5].

As an example, the testing of human gait is detailed hereafter.

4.2 Testing of human gait in vitro as an example

4.2.1 Materials, size of contacting parts and set-up

The dual mobility cup and the hip implant materials were delivered by SERF. The stem was a cemented one, Profil-1, the metal back (the metallic insert) was Novae-1 55 mm, the cup/insert was an ultrahigh molecular weight polyethylene (UHMWPE) with a diameter of 55–0.5 mm. The femoral head had a diameter of 22.2 mm. The femoral stem was cemented with cement Palacos R40 from Schering Plough. The fixing and positioning steps were managed by orthopedic surgeons. Details on manufacturing, such as polishing steps, machining processes, dimensions, etc., have to be available from the hip joint manufacturer.

Standard walking conditions can be reproduced by using standard ISO 14242 [6–8], e.g., using the hip walking simulator MTS 858 Mini Bionix II. First of all, such a machine can be used in traction/compression mode. The calibration of a servohydraulic machine is done as usual: fixing a setpoint value and adjusting the measured value from the transducer through the investigated range of normal loads. Following this pristine calibration step, some adduction/abduction and flexion/extension steps can be added. The applied load range can be varied stepwise from 300 N (compression) up to 5 kN.

The performance of the pressure-sensitive film based on ink appearance needs to be tested and the recommendation is that an acceptable reproducibility defined as getting some measured values from various operators with a minimum difference through five measurements, i.e., roughly and from the practice less than 20% of the difference is achieved. The Tekscan film was performed better than the pressure-sensitive film from Fuji [9–11]. One may easily notice that the protocol will influence the measured/monitored value. One may imagine that a plastic film under 20 MPa will suffer from creep. Practically it means that the deformation of every sensor is going to change according to the duration of the measurement. In polymer science (more generally physics), this phenomenon is the relaxation time. Calibration of the sensors and the software is required. Some key points are:

- the relaxation time is the resting period of the transducer from loading. In this work it was up to 60 s;
- during the measurement step, the value recorded after 10 s was considered;
- the transducer sensibility is to be considered since it affects the accuracy;
- the number of calibration data points is to be considered. In this work, it was set at nine points.

In the experimental protocol, the hip implant was fixed in the hip walking simulator according to ISO 14242-1. The insertion of the film between the UHMWPE insert and the concave face of the metallic metal back is shown in Fig. 4.5B.

The metal back is inserted in the acetabulum, actual case; it is fixed on the top part of the hip walking simulator, Fig. 4.5A. The considered Metal Back was manufactured from 316L, stainless steel, rod. The concave face of the Metal Back is in contact with the convex face, the external one, of the UHMWPE cup. In Fig. 4.5B, the pressure-sensitive film is inserted, like a sandwich, between the UHMWPE cup and the metal back.

The position of the UHMWPE insert has an impact on the stress measurements [12], Fig. 4.6. This experimental fact originates from the finding on explants, that a wear zone is located around the latitude of 66 degrees [4] from the Apex, latitude 0 degree of the cup [13].

Finally, about the experimental protocol, one may consider a lubricant mimicking some articular lubricant, during the loading phase. Hereto, some liquid droplets consisting of bovine serum were inserted between the pressure-sensitive film and the insert (UHMWPE cup) and between the metal back and the film. As mentioned before, a thin plastic film made of poly(vinylidene chloride) was used to wrap the sandwich made up by UHMWPE cup + film + metal back. Not usually well described, the clearance between head and cup is a key point to consider.

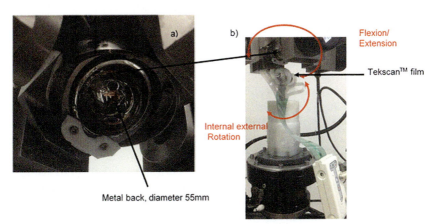

Fig. 4.5 (A) Metal back inserted in the hip walking simulator; (B) Film inserted between the dual mobility insert and the Metal back.

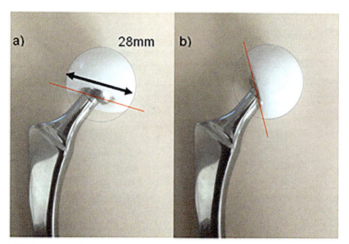

Fig.4.6 (A) Horizontal and (B) vertical positioning of the UHMWPE cup, insert.

From the physical point of view, adsorption on every surface in contact (Van der Waals forces and Helmholtz double layer) may occur in a liquid as bovine serum. The entire understanding (from the physical point of view) is not quite easy; however, from the experimental point of view, one may consider reproducibility, the statistical one, as a requirement. The statistical approach of every experimental measurement (five values measured of every experimental condition) was based on R software (https://www.r-project.org). The serials were compared based on the ANOVA test (http://math.univ-lyon1.fr/~duheille/MASS42_anova.pdf).

4.2.2 Some results on the maximum pressure recoded by the pressure-sensitive film

One pays only attention to the final results on the maximum contact stresses in different locations of the contact area. This maximum contact pressure appears as a one spot, gray color, on the pressure film and it is changing during the hip walking simulator movement). The maximum pressure (stress) at different locations is shown in Fig. 4.7A. As mentioned in reference [13], the study of explants revealed that the location with the deepest wear scar is around 66 degrees. From Fig. 4.7A, it appears that indeed the location of the maximum contact pressure corresponds to the more wearable zone. The fitting between the wear on explants, after approximately 10 years of using as implants inside patients, and the mapping of maximum pressure during a test of 10 s with a pressure-sensitive film is a most promising finding, Fig. 4.7B. The maximum contact stress is not always located at the same position during the movement of the metal back, usual test through hip walking simulator: flexion/extension, internal/external rotation, and cyclic normal load, Fig. 4.7A. Notwithstanding the worn zones are related to applying the maximum stresses. The assembly was not quite the same as the one in the human body. The maximum is located around 40 degrees from the Apex, white lane. The Dual mobility is changing its position during the gait

a)

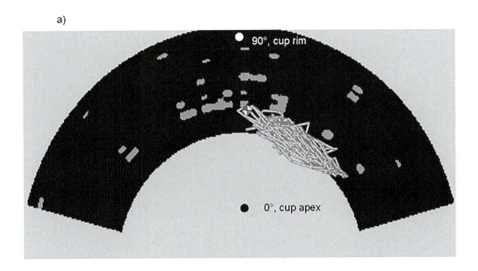

b)

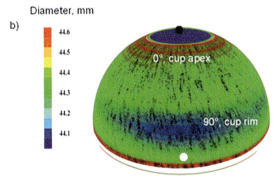

Fig. 4.7 (A) Image of the pressure sensitive film recorded during 10 s on the hip walking simulator movement obtained by Iscan^M software, trial 2; (B) Worn cup image from explant after 10 years, the *blue zone* represents the deepest wear zone (the lowest diameter); the *red zone* is indicated the no wear zone (the highest diameter).

as shown from the experimental data, Fig. 4.7B. Notwithstanding the maximum contact pressure is still related to the highest wear zone. In Fig. 4.7A, the gray dots are related to the shearing of the pressure film between the metal back and the UHMWPE cup. They may be considered as artifacts. In the Fig. 4.7B, an image corresponding to an explant is highlighting. The scale is roughly respected on both images, Fig. 4.7A and B. The main key point of this comparison is to show a proof of concept: by measuring live the maximum contact pressure, the more wearable zone may be reliable.

Some measurements of the load thanks to the transducer were practiced according to two positions of the cup (horizontal or vertical, see Fig. 4.6) in dry and wet conditions, Fig. 4.8. In the middle, the legend is as followed: vertical, with/dry. It means the vertical position of the cup with the film and in dry conditions. The decreasing way of

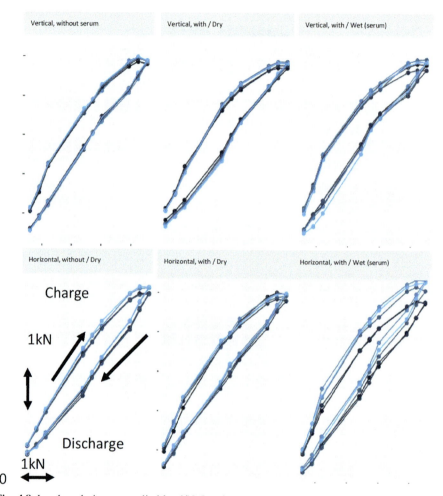

Fig. 4.8 Load evolution vs applied load/N. Loads were measured thanks to the transducer.

the load that is called discharge highlights that a remaining force is measured. The film is "prestressed." It is worth noting that the zero transducers is not reached immediately during the discharge phase. Five "round trip" curve has been considered corresponding to one condition: vertical, with/dry as an example. The highest discrepancy is reached concerning the horizontal orientation in wet serum, i.e., considering these values, the confidence interval is ±66 N for each normal load. The maximum of confidence interval is 5 kN, i.e., around 1%.

The minimum is 500 N, thus reaching more than 10% about the discrepancy. Two main conclusions are available from this measurements step:

- the first is that the protocol is realistic and it sounds robust with a gap of 10%;
- the second is that the serum medium is increasing the discrepancy of the measurements, as expected.

4.3 Applications in tribocorrosion

Some tribocorrosion issues, like contact mechanics/materials/solution/surface, etc., require a close look at any aspect of the specific case study on which the scientists are focusing on. Concerning the suggested case study of a dual mobility cup in orthopedics and the in situ of stresses, four points have to be emphasized, namely:

– cup positioning/samples positioning influence;
– liquid influence/contact pressure influence;
– the clearance impact;
– tribocorrosion and pressure-mapping film.

4.3.1 Cup positioning/samples positioning influence

The zero (absolute reference of the transducer) was different according to the test conditions. The stress mapping varies depending on the cup position [4]. The differences are around 20%. This experimental fact should be originally one way to explain why the wear zones are not regular on explants [13]. On the explants, a worn area can be noticed with these irregularities. By the way, the expected mechanism is reinforced through the experimental results from [4]. Two configurations can be considered, namely called horizontal and vertical ones. During the actual cup displacement, these positions are well-identified and highlighted.

4.3.2 Liquid influence/contact pressure influence

First of all, these results in solution are submitted to the protective film influence. The film is made of poly(vinylidene chloride-PVC). It is commonly used in the packaging industry. It has no huge effect on the force measurements recorded by the pressure-mapping film. Some measurements were investigated in the liquid environment without the pressure-mapping plastic film [4]; it was not working fine: bad reproducibility. Due to swelling of the plastic film (PVC) surrounding material, the measurements were changed between two tests, i.e. according to the time; that is the reason why the film is necessary to investigate under wet conditions to protect the pressure-mapping film. In a liquid environment, the presence of proteins involves the adsorption inside the insert; thus that is involving a Helmholtz double layer.

Moreover, due to the elaborated "sandwich" structure consisting of, insert/protective film/pressure-mapping film/protective film, the interfaces are multiple and the proteins adsorption may occur on any interface. A bearing effect was occurring in a liquid environment. Surely, it is happening but it may be due to proteins, due to the protective film, or the pressure-mapping film. One may think, in the beginning, that it was due to protein adsorption but finally the film folds took into account the surface change measurements.

About the fatigue of the materials, the pressure-mapping film description testified that fatigue did not affect measurements. Under the fatigue of the film, one may think about the polymeric one that covers the electric wires. With the covering system, it is not obvious to see any degradation. However, the practitioners noticed a zero change with time. The triggering was not due to an evident experimental cause. Finally, the folds and the fatigue of the sensors seem to play a major role compared to any physical phenomenon like adsorption.

4.3.3 The clearance impact of any contact

The clearance is well known in orthopedic implants and is defined as the space between the ultrahigh molecular weight polyethylene (UHMWPE) insert and the head (Metallic material, 316L Stainless Steel). Moreover, related to our case study, the clearance is considered as the space between the insert and the metal back. As emphasized by Rieker et al. [14], the clearance has a huge impact on the kinematic of any hip implant system. Moreover, Li et al. [15] showed by modeling and numerical simulation that the clearance is a key point in any measurement of the stresses on hip joints (natural or artificial). Due to that fact, an additional gap by machining was intentionally maintained to insert the pressure-mapping film. Thus a gap, i.e., clearance, was created between surfaces in contact. For implants company, this parameter is quite hard to control due to "machining" know-how and "usage" issues. A specific roughness is generated. The grooves generated by machining induce stresses and some deformation of the material. The physical key point is the thickness of the bearing liquid inside the grooves acting as an adsorbed layer between the materials in contact. This liquid accumulation promotes the relative displacement of the two materials in contact. This fact has a protective effect about soliciting the material by applying some stresses.

Thus, one may think about modeling the physical situation. Rieker et al. [14] described the methodology needed to get enough results. As usual, it is dealing with back and forward approaches between experimental investigations and modeling. Moreover, each company has its own "machining secret" to achieve the best clearance efficiency and related lubrication effect. The squeaking effect in Ceramic on Ceramic (CoC) junctions was reported in [16] and the last advancements were discussed. The statements and mechanisms were pedagogically shown through Fig. 4.8 [16].

4.3.4 Tribocorrosion and pressure-mapping film

A concern in tribocorrosion is to acquire a stress cartography to acquire some information on the surface solicitations. In a complex environment, the experimental use of a pressure-mapping film is a right way to do that job. However, this can be tricky as many pitfalls can arise when using pressure-mapping films. Finite element modeling is a tool that predicts stress mapping, but it does not predict actual stresses due either to roughness and/or to the surface state. Anyone who practises finite element modeling is faced with multiscale limitations.

By overviewing the stress mapping techniques used to measure certain mechanical constrain, some bias may be triggered due to measurement tools. One may wonder either the instrument influences the targeted phenomenon. Every experiment has the same statement. In some tribocorrosion investigations, the mechanical part needs to consider this issue. As an example, the measured tangential load has to be realistic according to the imposed normal load and the fixing process. Moreover, about the corrosion part, which is related to open circuit potential, the current density at a given potential, etc., the mechanical part may influence the electrochemical one. To obtain realistic measurements of any electrochemical data, the influence of mechanics on the electrochemical data has to be checked from the physical point of view. As an example, checking the material with the lowest Young's modulus in any assembly should be the safer attitude because the material with the largest deformation will determine the mechanical deformation of the assembly. Through an X–Y spatial resolution of 0.62 mm the roughness impact, till some microns to some nanometers, is not obvious to highlight. The electrochemical degradation, i.e., corrosion, is occurring at the atomic scale, $10^{-12}/10^{-10}$ m. Roughly eight orders of magnitude are separating the mechanical and electrochemical scales. The tribocorrosion topic is facing every time the multiscale issue challenge.

For illustrating that point, intergranular corrosion is promoting thanks to mechanical stresses, i.e., stress corrosion cracking. The physical facts are multiple: the mechanical stresses are taking into account the corrosive dissolution; they are enhancing by a factor of 100 the corrosion, through the wear volume. In 2002, some new directions about mapping tribocorrosion were suggested by Stack [17]. As evoked below, tribocorrosion is the multifactorial issue: mechanical-tribological-corrosive. From 2002, some scientific tools were developing for addressing any practical issue. Thus it is worth noting that some advancements were achieved and the pressure mapping is considered as a macroscopic help to predict the affected and critical zones by tribocorrosion and tribology in any assembly.

Finally, as a conclusive statement, this work showed a kind of proof of concept: practicing some stresses mapping in a liquid environment between materials in contact, under some electrochemical monitoring.

4.4 Conclusions

The experimental challenge presented in this work is dedicated to monitoring the contact stresses between materials in contact. The case study presented is the dual mobility hip implant.

The questioning on the actual area measured by the pressure-mapping film is presented as an example. However, the maximum contact stress area is highlighted by the experimental data obtained using the pressure-mapping film in a concave situation, i.e., the external side of the dual mobility insert.

The bearing effect related to the serum was highlighted thanks to the experimental procedure. The relevant issue is that it makes sense with the physical statement. Thus one can think of optimizing the pressure-mapping film. This development may be considered by the manufacturers.

Joining the development of the pressure-mapping film technology mentioned in this work and a tribocorrosion issue may inspire novel tribocorrosion investigations [18].

4.5 Notations

R the resistance (Ω)
ρ the resistivity (Ω m^2/m)
L length (m)
S area (m^2)

Acknowledgments

The authors are grateful to BPI France, through Serf Company for sponsoring the acquisition process of the film transducers. Some acknowledgments are dedicated to Dr. S. Boulat, Dr. A. Dangin and Pr. F. Farizon Department of orthopedic surgery, University Hospital Centre of Saint-Etienne-France, Ms. B. Bonnet from Mines Saint-Etienne-France and J.-L. Aurelle from SERF company Décines-France.

References

[1] E.M. Hasler, W. Herzog, G.H. Fick, Appropriateness of plane pressure-sensitive film calibration for contact stress measurements in articular joints, Clin. Biomech. 11 (1996) 358–360.
[2] R. Zdero, Z. Mahboob, H. Bougherara, Fujifilm measurements of interfacial contact area and stress in articulating joints, in: R. Zdero (Ed.), Experimental Methods in Orthopaedic Biomechanics, Academic Press, 2016, pp. 251–266 (Chapter 16).
[3] T. Chen, H. Wang, B. Innocenti, Tekscan measurements of interfacial contact area and stress in articulating joints, in: R. Zdero (Ed.), Experimental Methods in Orthopaedic Biomechanics, Academic Press, 2016, pp. 267–284 (Chapter 17).
[4] S. Boulat, B. Bonnet, J.-L. Aurelle, B. Bertrand, F. Farizon, J. Geringer, Mapping mechanical stresses on dual mobility cup, hip prosthesis: successful and promising issue, in: 2018-A-**2632**-ORS proceedings New Orleans-LA-USA, 2018.
[5] D. Wilson, C.A. Niosi, Q.A. Zhu, T.R. Oxland, D.R. Wilson, Accuracy and repeatability of a new method for measuring facet loads in the lumbar spine, J. Biomech. 39 (2006) 348–353.
[6] J. Uribe, J. Geringer, L. Gremillard, B. Reynard, Degradation of alumina and zirconia toughened alumina (ZTA) hip prostheses tested under micro separation conditions in a shock device, Tribol. Int. 63 (2013) 151–157.
[7] A. Perrichon, B. Reynard, L. Gremillard, J. Chevalier, F. Farizon, J. Geringer, Effects of in vitro shocks and hydrothermal degradation on wear of ceramic hip joints: towards better experimental simulation of in vivo aging, Tribol. Int. 100 (2016) 410–419.
[8] A. Perrichon, B. Reynard, L. Gremillard, J. Chevalier, F. Farizon, J. Geringer, A testing protocol combining shocks, hydrothermal aging and friction, applied to zirconia toughened alumina (ZTA) hip implants, J. Mech. Behav. Biomed. Mater. 65 (2017) 600–608.
[9] M.L. Harris, P. Morberg, W.J.-M. Bruce, W.R. Walsh, An improved method for measuring tibiofemoral contact areas in total knee arthroplasty: a comparison of K-scan sensor and Fuji film, J. Biomech. 32 (1999) 951–958.
[10] J.Z. Wu, W. Herzog, M. Epstein, Effects of Inserting a pressensor film into articular joints on the actual contact mechanics, J. Biomech. Eng. 120 (1998) 655–659.

[11] K.N. Bachus, A.L. DeMarco, K.T. Judd, D.S. Horwitz, D.S. Brodke, Measuring contact area, force, and pressure for bioengineering applications: using Fuji film and TekScan systems, Med. Eng. Phys. 28 (2006) 483–488.
[12] R.K. Korhonen, A. Koistinen, Y.T. Kontinent, S.S. Santavirta, R. Lappalainen, The effect of geometry and abduction angle on the stresses in cemented UHMWPE acetabular cups-finite element simulations and experimental tests, Biomed. Eng. 4 (2005) 32.
[13] J. Geringer, B. Boyer, F. Farizon, Understanding the dual mobility concept for total hip arthroplasty. Investigations on a multiscale analysis-highlighting the role of arthrofibrosis, Wear 271 (2011) 2379–2385.
[14] C. Rieker, R. Schön, R. Konrad, G. Liebentritt, P. Geneppt, M. Shen, P. Roberts, P. Grigoris, Influence of the clearance on in-vitro tribology of large diameter metal-on-metal articulations pertaining to resurfacing hip implants, Orthop. Clin. N. Am. 36 (2005) 135–142.
[15] J. Li, X. Hua, Z. Jin, J. Fisher, R. Wilcox, Influence of clearance on the time-dependent performance of the hip following hemiarthroplasty: affinity element study with biphasic acetabular cartilage properties, Med. Eng. Phys. 36 (2014) 1449–1454.
[16] E. Askari, P. Flores, D. Dabirrahmani, R. Appleyard, A review of squeaking in ceramic total hip prostheses, Tribol. Int. 93 (2016) 239–256.
[17] M.M. Stack, Mapping tribo-corrosion processes in dry and in aqueous conditions: some new directions for the new millennium, Tribol. Int. 35 (2002) 681–689.
[18] J. Geringer, M.T. Mathew, M.A. Wimmer, D.D. Macdonald, Synergism Effects During Friction and Fretting Corrosion Experiments-Focusing on Biomaterials Used as Orthopedic Implants, Woodhead Publishing Series in Biomaterials, 2013, pp. 133–180 (Chapter 2).

Further reading

A. Ruggiero, R. D'Amato, E. Gomez, Experimental analysis of tribological behavior of UHMWPE against AISI420C and against Ti6Al4V alloy under dry and lubricated conditions, Tribol. Int. 92 (2015) 154–161.

A. Tudor, T. Laurian, V.M. Popescu, The effect of clearance and wear on the contact pressure of metal on polyethylene hip prostheses, Tribol. Int. 63 (2013) 158–168.

H.-Z. Zhang, Y.-F. Zhou, W.-P. Li, H. Luo, J.-T. Wu, C. Jiang, Z. Chen, J.-Y. Hou, R. Yang, B. Song, Z.-Z. Zhang, Tibiofemoral contact mechanics after horizontal or ripstop suture in inside-out and transtibial repair for meniscus radial tears in a porcine model, Arthroscopy 37 (2021) 932–940.

Parallel wear tests: The need for statistical analysis in tribology

E.P. Georgiou[a,b], D. Drees[b], T. Van der Donck[a], S. Economou[c], and Jean-Pierre Celis[d]
[a]Dept. Materials Engineering (MTM), KU Leuven, Leuven, Belgium, [b]Falex Tribology NV, Rotselaar, Belgium, [c]Ecoinnovations, Halandri, Greece, [d]KU Leuven, Leuven, Belgium

5.1 Introduction: Lab testing approach to industrial issues of tribocorrosion

When a laboratory test campaign involves a large number of samples, the operator is often confronted with multiple choices having considerable repercussions on both its duration and final cost. The strategy to shorten invidual tests to reduce the total testing time for a project involves some risks. Short and simplified lab tests may have a poor correlation with the actual industrial applications, as they do not generate the same mechanisms.

Moreover, whatever choice is made for conducting laboratory tests, there is an additional unavoidable problem, namely the reproducibility of these tests. Indeed, the number of repeat tests determines the confidence level that can be used to draw a conclusion. This opens the door to a more practical experimental approach, which is known as parallel testing. With this approach, it is possible to study various conditions, for example in more or less aggressive environments such as tribocorrosion, and to have results which are sufficiently reliable and representative of the statistical behavior of an industrial system under field usage conditions. This statistical approach is needed to analyze and interpret tribological results and to evaluate the "variance" of a tribo-system. In statistics and probability theory, variance is a measure of the dispersion of the values of a sample or the probability distribution. The variance of a series of data indicates how the spread of data is. The closer the variance is to zero, the closer the data are to each other.

The use of this tool is treated in this chapter by describing in a first part the statistical tool itself and what we can expect from it. The second part of the chapter focuses on the presentation of results obtained during wear test campaigns conducted in environments and situations as dissimilar and distant as can be biology, mechanical engineering, and aeronautics.

Even though tribocorrosion is a frequently found issue in the industrial field [1], the electrochemical and mechanical interactions are not yet fully understood. Up-to-date, various electrochemical measurements and sliding tests protocols are proposed to get a better insight in the complex interaction between mechanical and electrochemical processes [2–4]. However, the behavior of tribosystems immersed in aqueous solutions cannot be predicted by simply collating data from electrochemical studies carried out without applying friction, and sliding tests performed under dry conditions [5]. Indeed, friction affects the reactivity of metallic surfaces, while electrochemical processes can modify the composition and mechanical properties of surfaces. To make things even worse, wear unlike friction can be quite unrepeatable, especially when the tests are performed under low contact pressures that simulate the infield application. A typical example can be seen in Fig. 5.1, where the significant spread of wear can be found between ten independent measurements on bare and DLC-coated steel samples. Notice that the spread on wear (k factor) increases as the wear phenomena progress.

Wear simulation is even more challenging in the case of adhesive wear, since it is less homogeneous. To obtain a realistic simulation in adhesive wear conditions moderate loads must be used. This leads to low wear rates, which in turn are difficult to

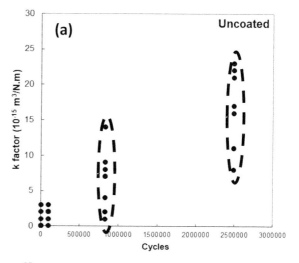

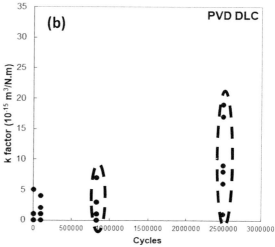

Fig. 5.1 Evolution of wear (k factor) as a function of sliding cycles for uncoated and PVD DLC coated carbon steel sliding against 10 mm Ø Teflon pins (applied load of 21 N, reciprocating sliding with a stroke of 10 mm, 1.25 Hz frequency, dry environment). Notice the spread of values between the independent measurements on different samples.

measure (by weight loss or profilometric volume loss). Adhesive wear is less repeatable and large variation on wear damage (>200%) are not unusual. In addition, alloys, and composites, have an inherent variability in friction and wear due to their complex microstructure and heterogenic structural features. For all of the above reasons, it is necessary to run longer tests, repeated tests, and tests with different amounts of cycles to draw safe conclusions on the wear resistance of materials under adhesive conditions. The only economically viable solution is to develop and use parallel wear generators that allow the simultaneous testing of many components.

In any case, the question that arises is: "How can the wear be measured in a repeatable way, especially under tribocorrosion conditions where the synergism between corrosion and mechanical loading can further complicate and accelerate the occurring phenomena." Solving this question and understanding the mechanisms that are at play, will lead to a better understanding which will help potentially in turning a negative synergy into a positive one. Thus, there is a growing need to study the wear under tribocorrosion conditions in a repeatable way.

5.2 Need for statistics to analyze wear

To tackle the unrepeatable and unpredictable "pattern" met in wear measurements and to obtain data with high confidence levels, statistical analysis is essential. Performing statistics is the only way to ensure that the wear data are interpreted correctly and that apparent relationships are meaningful (or "significant") and not simply chance occurrences.

For the analysis and ranking of wear different numerical values can be employed. The most commonly used one is the average value, which is defined as the sum of all wear measurements (e.g., wear width, depth, area, or volume) divided by the total number of measurements, as indicated in Fig. 5.2. However, one of the major drawbacks of using average values as a reference, is that the outliers are also taken into account. Especially in tribological measurements, outliers can occur due to e.g., misalignment of the contact, experimental errors, inhomogeneity of the material etc. Thus, depending on the frequency of appearance of these outliers, one has to evaluate whether they indicate variations in the material or they are simply an experimental error which should not be taken into account. In the latter case, the median value is a better value to be used, as it attaches less importance to outliers. Another option is to use the average value but at the same time expresses the spread of wear measurements by displaying standard deviation.

To characterize a single material couple, under one set of test conditions, usually requires at least 3 repeats at each duration. With a single-station test machine and tests that may take weeks to perform, it is economically and practically impossible to generate meaningful statistical data. A typical example of the importance of statistical analysis of wear is given in Fig. 5.3. If one performs 4 tests (represented by the values indicated by the red color) you will get a similar average value (*red line* in Fig. 5.3), as by performing all 15 repeats (*blue line* in Fig. 5.3). However, the 99% confidence levels for each sample size (4 vs. 15) will be extremely different. 10 repeats is

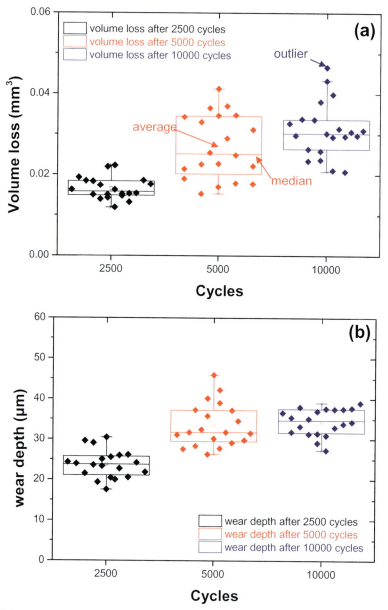

Fig. 5.2 Box plots showing spread of values average, median, outliers etc. The values presented illustrate the evolution of wear (expressed by the wear depth and volume of formed track) of a commercially supplied tool steel sliding against 6 mm Ø alumina pins (1 bar, 100 rpm, dry environment).

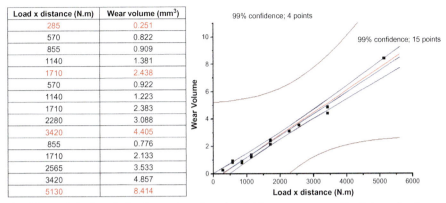

Fig. 5.3 Statistical analysis of wear based on 4 and 15 measurements. Confidence levels.

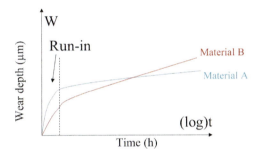

Fig. 5.4 Schematic evolution of wear and effect of run-in.

considered an optimum balance of obtaining enough data to perform statistical analysis with high confidence and saving time in sample preparation and analysis.

Another factor that should be considered in this approach is that tests are performed under low contact pressures to simulate "actual" contact conditions, like in an application, also require a long test time. This is extremely important if we consider that in the majority of tribological tests, there is a run-in period. If the test is too short (to save on time and cost), there is a risk that misleading conclusions will be drawn. A typical example is given in Fig. 5.4. If a short test is performed, then material A (indicated by the blue line) will appear to have lower wear resistance than material B (indicated by the red line), which is not the case for the longer duration. Such phenomena can occur for example due to removal of initial roughness, formation of a tribo-layer etc.

In addition, to evaluate how significant are the measured differences between various sample groups (e.g., wear measurements per material) analysis of variance (ANOVA) can be performed [6]. This statistical method allows to compare the mean values of three or more datasets, by hypothesizing that there is no difference among their mean values. This is achieved by comparing the mean squares of these datasets [6]. If there is a significant difference, then at least two datasets are different from each other. However, ANOVA analysis does not identify which are those. For this reason, post analysis is performed by pairwise comparisons of the datasets [7], in order to have a clear view on which pairs show a significant difference. One of the most common methods for pairwise comparison is Tukey tests [8]. These tests compare the absolute value of the difference between pairs of means and dividing it by the standard error of the mean (SE) [9]. The aim is to find out whether the "total variance" is due to the fluctuation of the values within each group (as shown in the example of Fig. 5.1) or between the groups (as shown in the example of Fig. 5.2). In particular, the larger the ratio of the fluctuation between groups to the fluctuation within each group (F = between groups/within groups), the more likely it is that these groups have different mean values. Among the different types of ANOVA analysis [10], one-way and two-way are commonly used in Tribological applications [11, 12]. The main differences between them are as follows:

- One way ANOVA tests the equality of three or more means simultaneously, whereas two-way ANOVA evaluates the interrelationship between factors, influencing variable.
- There is only one factor or independent variable in one way ANOVA (e.g., testing materials with different compositions, variable: composition), whereas in the case of two-way ANOVA there are two independent variables (e.g., testing materials with different compositions and at different loads).
- One-way ANOVA compares three or more levels (conditions) of one factor. On the other hand, two-way ANOVA compares the effect of multiple levels of two factors.
- In one-way ANOVA, the number of observations need not be same in each group, whereas it should be same in the case of two-way ANOVA.

To get a better understanding of ANOVA statistical analysis, an example is presented on how to apply it to evaluate wear. In Fig. 5.2, the effect of sliding cycles on the wear of a material is shown. From a first look differences between the box plots and their mean values can be seen. However, a question that arises is how significant are these differences. To analyze this, as a first step one-way ANOVA can be used since in this example there is only variable, the sliding cycles. Nowadays, ANOVA statistical analysis is implemented in multiple software packages such as MATLAB, Statistica, Origin, SPSS, and Excel. For this example Origin was used.

By performing one-way ANOVA the probability value (p-value) is calculated. This value tests the hypothesis that identical mean values are obtained from all examined groups. Subsequently, p-value is compared to a significant level, which is defined by the user. For example a significance level of 0.05 (or 95% confidence) indicates a 5% risk of concluding that "a difference exists when there is no actual difference." If the p-value is larger than the significance level, then the mean values of the compared groups are similar. On the other hand, a lower p-value than the significance level, indicates that at least one of the groups is significantly different. For Fig. 5.2, a p-value of $\approx 1.8 \times 10^{-12}$ was calculated, which is much lower than the 0.05 significance level. This confirms the initial observation that there is a difference higher than 5% in the wear track width between the measured sliding cycles.

However, this analysis did not provide with any information on which group/s had the highest difference (one or multiple?). For this reason a post-hoc test, such as Tukey's range test, is needed. Tukey's test performs a comparison of the mean values between each pair of examined groups (in our example: 10,000 cycles vs. 50,000 cycles, 50,000 cycles vs. 100,000 cycles and 10,000 cycles vs. 100,000 cycles) and

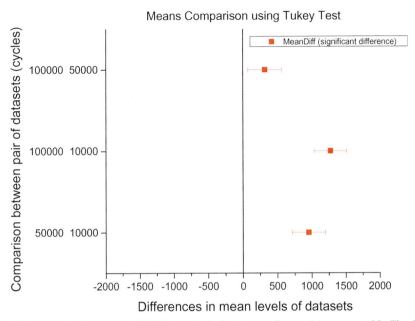

Fig. 5.5 Example of one-way ANOVA analysis for the set of wear data presented in Fig. 5.2.

evaluates whether their differences are higher than the standard deviation (fluctuation of measurements within each condition). The resulting outcome with a 95% confidence level (0.05 significance level) can be seen in Fig. 5.5. Based on this statistical representation, the above groups (wear for 10,000, 50,000, and 100,000 number of cycles) are compared in pairs and if their differences (depending on average and spread of values) are not significant, then they should cross the 0 line. Vice versa if they do not cross the 0 line their differences are significant. In this case, it is evident that all differences are significant, however the further away that the group is from the 0 line the higher the significance. Indeed, as expected the highest wear difference is expected between the 10,000 and 100,000 cycles, whereas what is also very interesting is that there is higher wear in the running-in period and therefore the significance is higher between 10,000 and 50,000 cycles than between 50,000 and 100,000 where steady state conditions are reached.

The need to perform statistical analysis to evaluate wear is unquestionable and this can be obtained only by virtue of parallel testing. This is becoming a key point in the automotive, aerospace, ceramics processing industry, medical implants, and oil & gas. These sectors need to know in advance product reliability in the field, to increase passengers and missions' safety (aerospace), patients' quality of life and to reduce their products' environmental footprint. In addition, in order to stay competitive in worldwide markets, they need to continuously bring into the market improved advanced and innovative products. This leads to a high demand for precise, statistically viable experimental methods that can be used to evaluate new advanced materials at low cost and within short times.

5.3 State-of-the-art on current wear generators

Rarely do laboratory wear test units simulate a tribocontact that exists in an industrial application. Typically ball-on-flat geometry is used, contrary to reality where area or line contacts are usually found, Fig. 5.6. Point contacts generate much higher initial contact pressures. For example, when applying just 1 N load with a 10-mm diameter steel ball on steel plate a Hertzian contact pressure of 500 MPa is already generated. This value is more than double of typical contact pressures in the automotive industry (100–200 MPa). Furthermore, a machine component will be designed for a lifetime of more than 2000 h with minor tolerance loss. Assuming a maximum acceptable wear depth of 2 μm, this calculates into a wear rate of 1 nm/h. However, in the majority of lab tests, some μm of wear are generated within the hour, so a 1000-fold accelerated wear mechanism is happening. The risk of such an accelerated wear test is clear: the wear mechanism may no longer correlate with the actual application so the test is not relevant. The realistic low wear rates need to be simulated on lab scale, so long tests are inevitable. For these reasons, test setups that can perform in parallel multiple wear tests at contact conditions that simulate actual industrial applications were developed. In addition, wear measurements (wear depth, volume, weight loss etc.) are known to display a notoriously large variation, as confirmed by the well-defined VAMAS studies in the 1980s [13, 14]. A typical example is presented in REF [13], where a repeatability (within one laboratory) of wear of ±14% and reproducibility of ±38% were found. As a result, it is necessary to perform enough repeat tests, so as to evaluate whether the results are significantly different by statistical analysis. This highlights the practical contradiction of wear measurements. On the one hand, realistic wear tests will generate low wear rates for many applications and require long tests (days to weeks), whereas on the other hand, repeat test results (ideally 3 to 5, or even 10) should be produced in an economically practical time frame.

The Falex Miller tester (Fig. 5.7) is used to investigate in an aqueous medium the interaction between engineering materials and abrasive particles, as in a slurry. It was standardized in the 1960–1970s culminating in the ASTM G75 standard [15]. This tester provides the rate of mass-loss of a specimen at a specific time on the cumulative abrasion-corrosion time curve, converted to volume or thickness loss rate. In particular, from this test, the Miller and Slurry Abrasion Response (SAR) numbers can be extracted. The Miller number is an index of the relative abrasivity of slurries [16]. Its primary purpose is to rank the abrasivity of slurries against a standardized reference

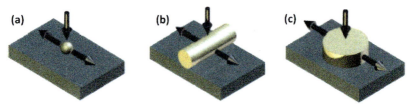

Fig. 5.6 Tribological sliding contact conditions: (A) point, (B) line and (C) area.

Fig. 5.7 Falex Miller tester.

material. A higher Miller Number indicates more wear damage on the specimen or a more abrasive slurry. A detailed analysis on how these numbers are calculated will be given in Section 5.4.2.

Even though this method is quite an old technique, research protocols are still being developed with the incorporation of a galvanostat/potentiostat for simultaneous electrochemical measurements, as in sliding tribocorrosion protocols. The evolution of wear can be observed, and repeatable results are obtained thanks to special design details of the equipment and the use of a consistent methodology.

A significant breakthrough in wear generator technology is the Circularly Translating Pin-On-Disk (CTPOD) [17]. A Super-CTPOD tester containing 50 or 100 stations is available [18] (Fig. 5.8). This tester can be set to reciprocating or unidirectional motion, use different counter materials, geometries and/or corrosive or lubricating environments. This type of wear generators is used extensively in the orthopedic field for the multidirectional motion of the polymeric pin, which avoids molecular orientation [19]. Unidirectional wear vectors are, in this case, not reliable for predicting the wear of e.g., prosthetic joints, as the obtained wear factors are typically two orders of magnitude lower.

This wear generator allows to evaluate material wear rates efficiently in both lubricated and unlubricated conditions. Each pin-disk contact can be lubricated individually, making the tester also suitable to compare simultaneously different environmental conditions, e.g., different corrosive environments.

To expand the test parameters to higher sliding speeds and pressures, another design of a 10-station rod-on-specimen configuration allowing to evaluate simultaneously the wear of 10 individual specimens is shown in Fig. 5.9. All pneumatically loaded stations apply the same pressure, but can be disengaged in pairs so that the number of test cycles can be easily modulated during a test. This allows efficient

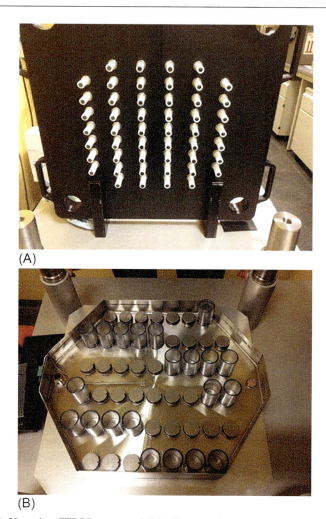

Fig. 5.8 (A) 50-station CTPOD tester and (B) tribocorrosion cups.

testing for measuring wear evolution. Contacts can be point (crossed cylinders), line (flat on cylinder) or area (conformal block on cylinder). Heating and drip-feed lubrication are integrated. A three-body sand abrasion option is also considered. This set-up allows for speed variation as well as heating (thermocouples are connected directly to the 10 samples) and lubrication. The possible contact geometries are presented in Fig. 5.10. A wide range of contact pressures can be achieved and different wear "scenarios" can be simulated. The number of test stations for parallel testing is reduced to ten in comparison to the 50–100 station CTPOD tester. This number of stations balance the need for statistics with after-test measurement effort. Ten data points per material provide enough statistical results and allow wear rate or evolution to be measured in an efficient way and with a relatively high confidence. This test method is

Parallel wear tests

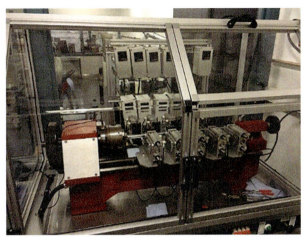

Fig. 5.9 Falex Tribology 10-station wear tester.

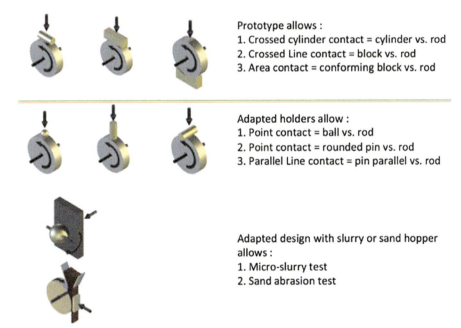

Prototype allows :
1. Crossed cylinder contact = cylinder vs. rod
2. Crossed Line contact = block vs. rod
3. Area contact = conforming block vs. rod

Adapted holders allow :
1. Point contact = ball vs. rod
2. Point contact = rounded pin vs. rod
3. Parallel Line contact = pin parallel vs. rod

Adapted design with slurry or sand hopper allows :
1. Micro-slurry test
2. Sand abrasion test

Fig. 5.10 Contacts that can be simulated with the Falex Tribology 10-station tester.

newly developed, so the repeatability and reproducibility are still under evaluation [20]. To the authors' knowledge, the repeatability of results (as with all the setups) strongly depends on the evaluated tribosystem and mainly on the structural and surface homogeneity of tested materials. Typical wear fluctuation with this setup are in the range of 10%.

5.4 Industrial examples on statistical wear analysis

5.4.1 Biomedical field

The need for parallel wear tests is unquestionable for the biomedical sector, since "reliability of materials" is of key importance. As there is an increasing demand to expand the lifetime of implants, new materials are being developed. The wear resistance of these materials should be measured in a systematic way and under conditions that simulate the actual contact. CTPOD wear generators have been found to generate similar wear rates and mechanisms [17], whereas they can also provide statistical evaluation of the material. They are the best alternative to full simulation tests, which are costly, time consuming and not always feasible in a development stage.

An example of the use of CTPOD wear generator to investigate implant materials is shown in Figs. 5.12 and 5.13 [17]. The wear resistance of nanostructured sprayed titanium cermets coatings and the influence of surface finishing on the wear of the tribosystem (coating vs ultrahigh molecular weight polyethylene UHMWPE) is shown for two variations of these coatings with different roughness were tested. As a reference a stellite benchmark was used. The tribosystem was immersed in calf serum and the wear of the UHMWPE was measured via weight loss measurements after a specific number of sliding cycles (Fig. 5.11). In this example, 15 samples of each coating/benchmark were tested for statistical analysis (Fig. 5.12).

It appears that the surface roughness of the coating had the most significant impact on the wear loss of the UHMWPE, as the asperities act as "cutting tools" that abrade the softer polymer. On the other hand, a smooth surface finish induced significantly less wear. Avoiding the formation of debris is extremely important, since they can cause inflammations in the body.

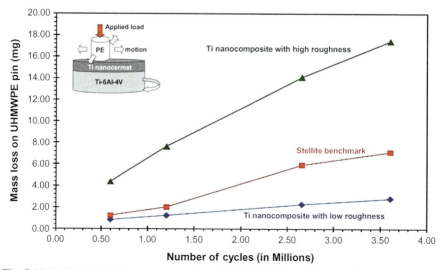

Fig. 5.11 Evolution of the average wear loss of UHMWPE pins sliding against nanostructured Ti cermets and stellite in 10% Newborn calf serum [19].

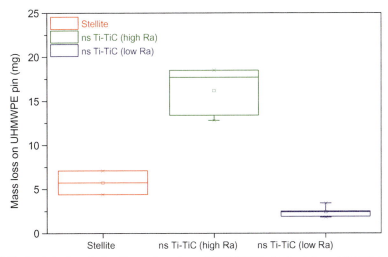

Fig. 5.12 Statistical analysis of wear loss data of UHMWPE obtained from CTPOD tests performed against nanostructured Ti cermets and stellite in 10% Newborn calf serum [10].

Having enough data points per test condition, allows a statistical analysis of the CTPOD wear data, in the form of 'box plots' (Fig. 5.12). This analysis shows that median and average values are similar, indicating good repeatability of test samples and process. Some outliers can easily be identified, they are possibly the result of the inhomogeneity of the surface. However, due to the large amount of data that is produced simultaneously, their influence on the average values is minimal. As a result, a confident and clear ranking between cermet coatings and the stellite benchmark can be made: the lower the roughness of Ti nanocermets, the smaller the damage on UHMWPE pins. What is also interesting is the spread of box plots. For example, the nanostructured Ti cremets with low roughness have the smallest spread, indicating better tribological homogeneity of the system. Finally, a clear ranking of the tribosystems can be seen, with high confidence levels.

5.4.2 Construction industry

Many components in the construction industry can be undergoing heavy abrasion in water. Some examples are mixers and scrapers in brick manufacturing, they undergo slurry abrasion from the clay mixtures. Such components typically have a lifetime of 75 hours, resulting in high maintenance costs, delays in the production line, production losses and production quality fluctuation.

To increase the lifetime of these components, new coatings are developed by various plating techniques such as electroless deposition, electrodeposition, thermal spraying, chemical vapor deposition (CVD), physical vapor deposition (PVD) and laser cladding. However, to evaluate the wear resistance and the reliability of these materials before putting to a field trial, pre-screening methods are needed to provide a first insight and selection. The closest simulation to this type of slurry abrasive mechanisms is the Falex Miller test. In the standardized ASTM G75 protocol [15], a mixture of abrasive sand particles with distilled water (in a ratio of 1/1 per weight) forms the abrasive slurry. The test can also be performed with customized or actual slurries. Then, a difference in the abrasivity is to be expected (Fig. 5.13A). The wear of a standard material against different slurries is called the Miller Number (MN), the

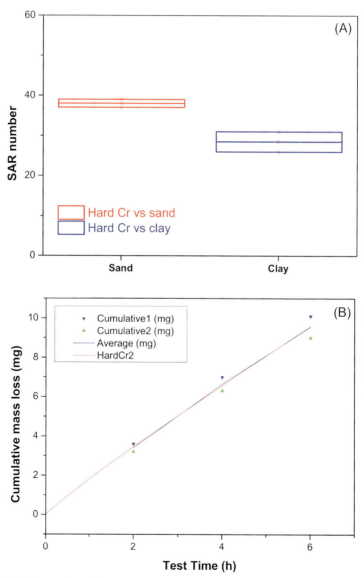

Fig. 5.13 (A) Comparison of the SAR numbers of a hard chrome coating tested in sand and clay slurries. (B) Evolution of mass loss for a pair of hard Cr coated blocks as a function of abrasion time in a clay slurry. These tests were performed according to ASTM G75.

wear of any material against any slurry is called the Slurry Abrasivity Response or SAR Number. The Miller and SAR numbers are calculated as follows. At first the weight loss of one pair of the same materials are plotted as a function of test time (green and blue points in Fig. 5.13B). Then, curve fitting of these data is performed with the use of the basic mathematical equation:

$$M = At^B \tag{5.1}$$

where M the cumulative mass loss, A the first curve fit coefficient, B the second curve fit coefficient and t the time.

The curve-fit line is indicated by the red line in Fig. 5.15B. The Miller and SAR numbers are indexes related to the mass loss of the materials at two hours into the test, which can be calculated by using the first derivative of Eq. (5.1) at 2 h, which is given by the following equation:

$$M_{nr} = A \times B \times t^{(B-1)} \tag{5.2}$$

The Miller and SAR numbers are subsequently calculated by Eqs. (5.3) and (5.4):

$$\text{Miller number} = M_{nr} \, (\text{in mg/h}) \times 18.18 \tag{5.3}$$

18.18 is a scale factor expressed in h/mg.

$$\text{Slurry Abrasivity Number} = MN \times (7.58/\text{specific gravity of material}) \tag{5.4}$$

The progressive slurry abrasivity tests, consisting of minimum 2 but up to 4 samples per test, are done for 2, 4 and 6 cumulative hours.

A comparison between hard Cr (HCr), reference WC material (REF), laser clad (LC), thermal sprayed (TS) and electrodeposited (ED) coatings is given in Fig. 5.14. This figure illustrates that existing plating methods can be used to replace existing hard metals (REF) or hard Cr coatings as they have a similar or lower SAR number. In this example, the TS coatings show a comparable resistance against slurry abrasion, particularly TS_E and TS_F.

5.4.3 Printing industry

One of the most common issues in the printing industry is the wear damage of printing rollers due to the synergistic effect of ink abrasivity and sliding motion between different components. For this reason, new coatings are being developed to improve the wear resistance of the rollers, whereas they also need to have an ink-repellent surface and a high resistance to standard cleaning agents used in printing technology.

These rollers have a specific texturing, namely the negative of the forms that need to be printed. This texture is getting smaller and smaller and requires higher precision to print with better resolution. A typical analysis of a standardized surface texture by confocal microscopy, to evaluate the roller durability and at the same time the texture

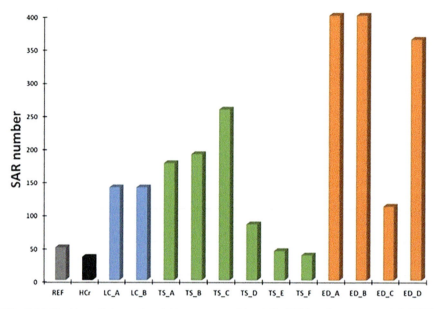

Fig. 5.14 Ranking of slurry abrasivity response of hard Cr (HCr), reference material (REF), laser clad (LC), thermal sprayed (TS) and electrodeposited (ED) coatings. A, B, C, D, E and F indicate a different composition and/or deposition conditions.

durability, is given in Fig. 5.15. Notice the microscopic rises and dips in the 2D profile (Fig. 5.18C). This texture holds the ink in tiny reservoirs and transfers it to the paper or printed material, if that texture wears to fast, resolution is lost and in the end, the ink cannot be transferred anymore.

To simulate the wear mechanism in these textured printing rollers and evaluate the wear loss with high confidence levels, the 10-station wear generator was used. To simulate the abrasive wear mechanism, alumina cylinders with dimensions 6 mm height × 6 mm Ø were used in a cross cylinder contact geometry. The aim is to simulate localized abrasion in a ceramic vs ceramic system. The applied load was 52 N (1 bar), generating a contact pressure of about 1.6 GPa, and the rotation speed was 50 rpm, which in this case is translated into a linear speed of 0.175 m/s. By performing ten parallel tests on just a single rod and at different intervals, the evolution of wear can be efficiently monitored with 95% confidence levels as described in Section 5.2 (Fig. 5.16). In this case, the wear damage was evaluated by confocal microscopy. One of the questions that arise, when studying such textured surfaces is which value is more important, namely either the mean wear depth (Fig. 5.16A), the max wear depth (Fig. 5.16B) or the volume loss (Fig. 5.16C). The interesting part is that a similar evolution is observed, however the 95% confidence levels appear to be significantly larger when volume loss is considered. This is because 3D volume analysis measurements is influenced strongly by topographical features such as waviness. Extracting multiple 2D profiles and analyzing the mean and the max wear depth of these is a better methodology for wear loss evaluation of textured surfaces.

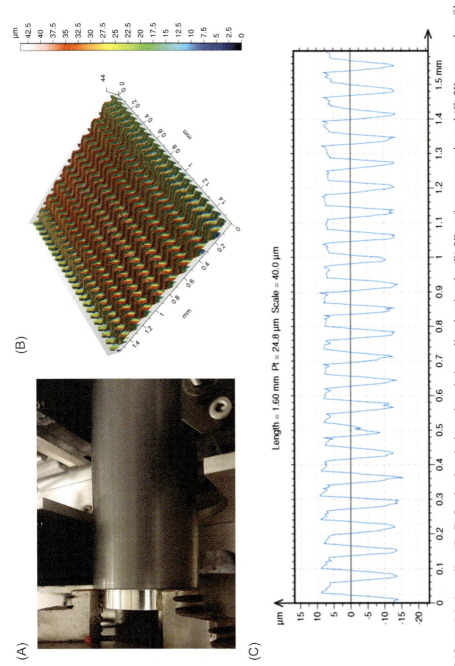

Fig. 5.15 (A) Printing roller. (B, C) Confocal analysis on the printing roller showing the (B) 3D surface topography and (C) 2D extracted profile (as indicated by line AB in Fig. 5.17).

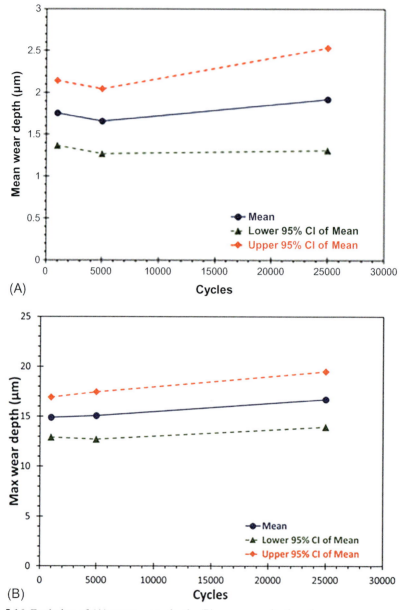

Fig. 5.16 Evolution of (A) mean wear depth, (B) max wear depth and

(Continued)

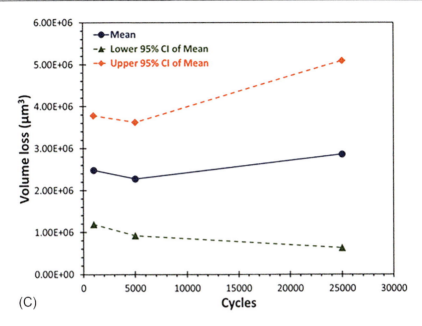

Fig.5.16, cont'd (C) volume loss of a printing roller alumina cylinders with dimensions 6 mm height × 6 mm Ø were used in a cross cylinder contact geometry. Blue line indicates the average (mean) values and the green and red lower and upper 95% confidence levels for these measurements.

5.4.4 Aeronautical field

There is an increasing demand for new materials in the aeronautical field. They should have high strength, low weight, corrosion, and wear resistance in a variety of environmental conditions. Among them, polymer composites are becoming a key candidate for such applications, since they have a high specific strength and strength to weight ratio. They may also be easier formed into complex shapes and have a higher corrosion and fatigue resistance than conventional aeronautical aluminum and titanium alloys. Already, polymeric composites are extensively used from luggage storage bins and floor panels up to the torsion box, pylon fairings and wing spoilers. However, they do have a lower wear resistance than lubricated metals or ceramics and can be sensitive to moisture. The combination of these two factors can accelerate their degradation, and thus requires an investigation into the tribocorrosion behavior of these materials.

An example of the evolution of wear of composite polymers for the aeronautical field, under tribocorrosive conditions, is shown in Fig. 5.17. Each point in this figure is an average of 25 independent measurements, performed with a CTPOD wear generator. From a first look it is evident that the composite (B) has a much higher wear resistance to tribocorrosion than composite (A). However, contamination from the environment can significantly change that balance. In particular, the presence of hard particles in the contact will have a substantial effect, the wear loss is accelerated due to

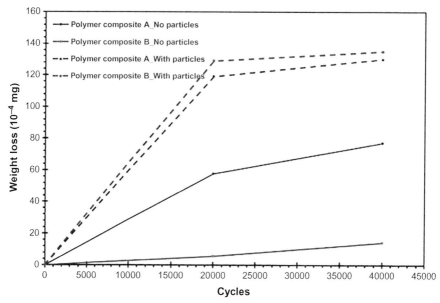

Fig. 5.17 Comparison of two polymeric composites in a tribocorrosion environment. The presence of debris in the contact have a significant influence on the generated wear damage.

three body abrasion mechanism and this influence makes the difference between the two composites almost disappear. The conclusion is, that the influence of the environment—and in this case the presence of particles—is the key factor in a correct tribocorrosion simulation.

5.4.5 Offshore installations

Also in offshore installations, wear under corrosive conditions is common. Examples are chain joints, drilling equipment, pipeline systems, flexible risers, mooring lines, turbine blades and other mechanical contacts. The failure of such components can lead to safety hazards and delay significantly production times. They are difficult to repair due to the complexity of the contact and the location of the installation. For this reason, new materials are being developed to withstand these harsh tribocorrosive conditions. Before putting these materials to the field test, prescreening is a cost-efficient way to reduce the risk. Any prescreening method needs to analyze the reliability of the material, under conditions that simulate the actual application (e.g., corrosion in seawater combined with low periodical stresses/reciprocating motion). The most efficient way to do that is by utilizing wear generators.

In the following example, three polymers named A, B and C were compared. Pins were prepared from existing materials, Fig. 5.18. In addition, they were manufactured in three different diameters to vary contact pressures from 1.7 MPa (low pressure L), 4 MPa (medium pressure M) and 8 MPa (high pressure L). The low contact pressures simulate again the real life conditions, but are very time-consuming (more than

Parallel wear tests

Fig. 5.18 Samples for CTPOD testing *(right)* manufactured out of actual component *(left).*

1.5 million cycles, 30 km distance covered and 1 month of testing time per sample needed). 316 stainless steel disks were the countermaterial, while the complete tribosystem was immersed in seawater. To obtain these contact pressures an applied load of 110 N was used with a flat-on-flat contact configuration, whereas the sliding speed was 25 mm/s.

The typical surface of the pins, after machining, shows the circular pattern of a turning operation. A 3D-confocal measurement of each type of pin (small, medium and large) is made prior to testing. One example of such a measurement is shown in Fig. 5.19A. After tribocorrosion tests for 1.5 million cycles at medium contact pressure, the initial surface topography has been completely removed and only grinding lines in the direction of the sliding motion remain (Fig. 5.19B). To quantify the wear, weight loss of each pin was measured with a precision of 0.1 mg.

To compare polymers, the wear factors were calculated by the following equation:

$$V = kFs \tag{5.5}$$

with V the volumetric wear (from mass loss and density), F the normal load and s the sliding distance.

This equation can be reworked by taking into account that nominal pressure $P = F/A$ with A the nominal contact area, and $h = V/A$ with h the height loss for a given volume, to the following form:

$$h = kPs \tag{5.6}$$

so that k can be calculated as

$$k = \frac{h}{Ps} \text{ or } k = \frac{V}{Fs} \tag{5.7}$$

In any case, the volumetric wear has to be calculated from mass loss/density.

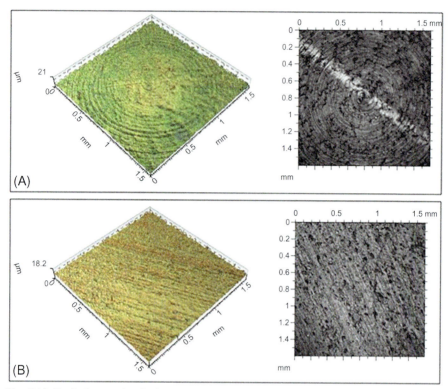

Fig. 5.19 Typical example of tribocorrosive damage on the polymer pins: (A) before and (B) after tribocorrosion testing.

From the total distance travelled for each pin, the area of the pin and the load upon it, this traditional wear factor can be calculated for each pin, and statistical evaluations of the measurement result can be easily made (Fig. 5.20). The summary of average wear factors reveals a major difference between polymer B on the one hand, and polymers A and C on the other hand (Fig. 5.20A). Also a statistically relevant comparison of polymers A and C could be made, with A showing to be a few % better than C. To evaluate the effect contact pressure on the wear loss of these materials one-way ANOVA statistical analysis (variable: contact pressure) was performed. An indicative example for polymer A with a 95% confidence level is shown in Fig. 5.20B. From Tukey's test analysis (see Section 5.2) it is evident that there is a "significant" effect of the contact pressure on the mean values of wear loss. This effect is more pronounced between the High and Low (H-ML) contact pressures, and least between High and Medium (H-M).

It is also noticeable, that the wear rate of material B depends largely on the area which is in contact with the steel disks achieved by increasing the contact pressure. Between materials A and C, that difference is not significant. All these test results, with different contact pressures and 5 repeats per condition, were obtained in one campaign of 4 weeks of testing, resulting in 50 data points of 30 km of accumulated sliding distance.

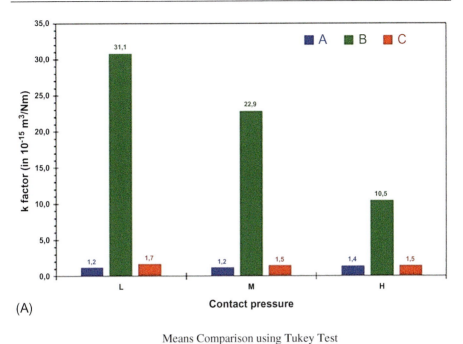
(A)

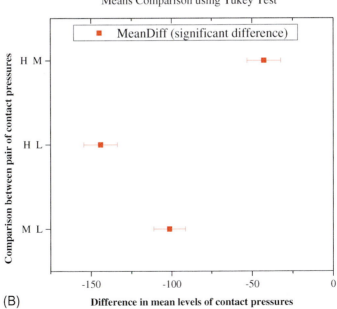
(B)

Fig. 5.20 (A) Comparison of average wear factor of polymers A, B and C under low L (1.7 MPa), medium M (4 MPa) and high H (8 MPa) contact pressures, in a seawater environment. (B) Tukey's test statistical analysis showing the significance of the effect of contact pressure on the mean wear loss of polymer A.

5.4.6 Industrial coatings

Understanding the interactions between surface corrosion and mechanical processes, and having efficient ways to measure them, will lead to the development of systems that can actually function in water lubrication. A typical example of potential water lubrication is given in Fig. 5.21. Ni-P nanocomposites attract industrial interest because of their mechanical and chemical characteristics and are considered a potential replacement for hard chrome coatings in some applications. The coefficient of friction under water 'lubrication' actually appears to be more stable and 20% lower than the same coating lubricated by a greasy film. So there seems to be no adverse effect of the water environment in terms of lubricated behavior.

Naturally, it is equally important to compare the wear damage synergy by water as an environment. In many cases, water lubrication can decrease the friction of a tribosystem when compared to dry conditions, but that does not mean that it also decreases the wear rate. In Fig. 5.22, the influence of water lubrication on the friction and wear of various industrial materials and coatings is summarized. For each coating or material, the value of friction or wear respectively is positioned in an X–Y graph, with the value in water lubricated conditions represented on the X-scale, and the dry conditions on the Y-scale. Any material that positions itself above the dividing line (where the coefficients are equal), would be a material that is positively influenced by the presence of the water lubrication. For instance, Fig. 5.22A shows that friction reduces when going from dry to water lubricated conditions for practically all materials (they are above the 1–1 line) but in terms or wear damage, the picture is more

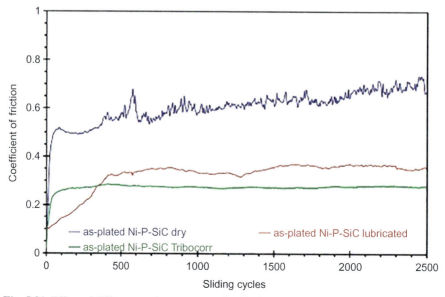

Fig. 5.21 Effect of different environments on the frictional behavior of Ni-P nanocomposites against corundum tested in a pin-on-disk configuration, under reciprocating sliding conditions (5N, 10 mm, 10 mm/s, 5 mm Ø alumina ball).

Parallel wear tests

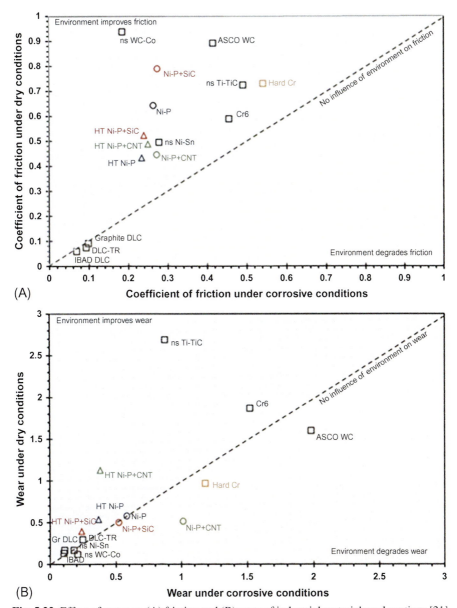

Fig. 5.22 Effect of water on (A) friction and (B) wear of industrial materials and coatings [21].

varied. Materials above the 1–1 line have a lower wear in water, those below this line show an increase in wear. This behavior strongly depends on the ability of the material to form protective or passive layers and on the influence of the friction mechanism on those layers. A methodology that allows to measure these properties efficiently and map different materials for comparison purposes, helps the industry to make better material selections for tribocorrosive conditions.

5.5 Summary

Tribocorrosion is an important issue in industry, as examples can be found in many fields and applications. Equipment can be adapted to testing materials in such conditions, to measure the synergism between electrochemical and mechanical phenomena. However, the evaluation of wear rates with high confidence levels requires the use of specialized equipment. This is because wear can display a high fluctuation and is unique for every tribo-system and test conditions. The main challenge is thus how to perform relevant experiments that simulate realistic conditions, and at the same time produce many repeats at an acceptable economic cost and duration. The only way to do this, is with the use multi-station wear generators. These machines perform multiple tests in parallel, allowing for a time and cost saving that makes statistical analysis economically possible.

Statistical analysis is one of the key factors in evaluating the reliability of materials under tribocorrosive conditions.

5.6 Notations

Symbol	Description	Unit
k factor	Wear factor	m^3/Nm
M	Cumulative mass loss	mg
Mnr	Miller number	–
SAR	Slurry Abrassivity number	–
P	Pressure	MPa
V	Volumetric wear	mm^3
F	Force	N
t	Time	h
A	Area	mm^2
s	Sliding distance	m
h	Height loss	mm

References

[1] P.R. Roberge, Corrosion Engineering: Principles and Practice, McGrawHill, 2008.
[2] P. Ponthiaux, F. Wenger, N. Diomidis, D. Dirk, J.-P. Celis, Protocol for testing the tribocorrosion behavior of passivating materials in sliding contacts, in: Corrosion From the Nanoscale to the Plant, 2009, p. 310.
[3] A. López-Ortega, J.L. Arana, R. Bayón, Tribocorrosion of passive materials: a review on test procedures and standards, International Journal of Corrosion (2018), 7345346.
[4] P. Ponthiaux, R. Bayon, F. Wenger, J.-P. Celis, Testing protocol for the study of bio-tribocorrosion, in: Bio-Tribocorrosion in Biomaterials and Medical Implants, vol. 60, 2013, pp. 372–394.
[5] N. Diomidis, J.-P. Celis, P. Ponthiaux, F. Wenger, A methodology for the assessment of the tribocorrosion of passivating metallic materials, Lubr. Sci. 21 (2009) 53–67.

[6] A. Gelman, Analysis of variance—Why it is more important than ever, Ann. Stat. 33 (2005) 1–53.
[7] H. Abdi, L.J. Williams, Encyclopedia of Research Design, Sage, Thousand Oaks, CA, 2010.
[8] K.V. Ramachandran, On the Tukey test for the equality of means and the Hartley test for the equality of variances, Ann. Math. Statist. 27 (1956) 825–831.
[9] J.W. Tukey, Exploratory Data Analysis, Addison-Wesley, Reading, 1977.
[10] R.A. Armstrong, F. Eperjesi, B. Gilmartin, The application of analysis of variance (ANOVA) to different experimental designs in optometry, Ophthal. Physiol. Opt. 22 (2002) 248–256.
[11] N. Koottathape, H. Takahashi, N. Iwasaki, M. Kanehira, W.J. Fingerd, Quantitative wear and wear damage analysis of composite resins in vitro, J. Mech. Behav. Biomed. 29 (2014) 508–516.
[12] G. Mitov, S.D. Heintze, S. Walz, K. Woll, F. Muecklich, P. Pospiech, Wear behavior of dental Y-TZP ceramic against natural enamel after different finishing procedures, Dent. Mater. 28 (2012) 909–918.
[13] E.A. Almond, M.G. Gee, Results from a U.K. Interlaboratory project on dry sliding wear, Wear 120 (1987) 101–116.
[14] H. Czichos, S. Becker, J. Lexow, Multilaboratory tribotesting: results from the Versailles advanced materials and standards programme on wear test methods, Wear 1114 (1987) 109–130.
[15] ASTM G75-15, Standard Test Method for Determination of Slurry Abrasivity (Miller Number) and Slurry Abrasion Response of Materials (SAR Number), 2015.
[16] J.E. Miller, F. Schmidt, Slurry Erosion: Uses, Applications, and Test Methods, American Society for Testing and Materials, 1987.
[17] V. Saikko, A multidirectional motion pin-on-disk wear test method for prosthetic joint materials, J. Biomed. Mater. Res. 41 (1998) 58–64.
[18] V. Saikko, A hip wear simulator with 100 test stations, Proc. Inst. Mech. Eng. H 219 (5) (2005) 309–318. ISSN 0954-4119.
[19] E.P. Georgiou, D. Drees, S. Dosta, P. Matteazzi, J. Kusinski, J.-P. Celis, Wear evaluation of nanostructured Ti cermets for joint reconstruction, Biotribology 11 (2017) 44–50.
[20] E.P. Georgiou, L.M. Lopes, M. De Bilde, D. Drees, How can we measure sliding wear in an efficient way? Wear 458–459 (2020) 203414.
[21] D. Drees, E. Georgiou, A. Zoikis-Karathanasis, T. Kosanovic Milickovic, I. Deligkiozi, J.-P. Celis, Nanostructured composite Ni-P electrodeposits as alternative to hard chrome coatings, in: STLE 2016, Las Vegas, Nevada, USA, 2016.

The use of the Pearson's correlation coefficients to identify mechanical-physical-chemical parameters controlling the tribocorrosion of metallic alloy

Vincent Vignal[a] and Halina Krawiec[b]
[a]Laboratoire Interdisciplinaire Carnot de Bourgogne (ICB), UMR 6303 CNRS—Université de Bourgogne-Franche-Comté, Cedex, France, [b]AGH—University of Science and Technology, Faculty of Foundry Engineering, Krakow, Poland

6.1 Introduction

Over the past decades, the use of electrochemical methods at the macro-scale (open-circuit potential (OCP) measurements, polarization curves, electrochemical impedance spectroscopy (EIS), noise analysis…) provided important information on the tribocorrosion mechanisms and on the kinetics of reactions. In this context, Landolt et al. [1] discussed the application of electrochemical methods to the study of tribocorrosion phenomenon. Ponthiaux et al. [2] discussed electrochemical techniques for studying the combined corrosion-wear degradation of materials in sliding contacts immersed in electrically conductive media. Other review articles have also been published [3, 4].

Surfaces subjected to sliding are usually highly heterogeneous (engineering materials). Quantitative information about the mechanical and electrochemical properties of individual phases, particles, defects, sites with stress gradients… is important data to develop realistic models describing tribocorrosion mechanisms and corrosion rates [5]. For roughly 20 years, the use of local electrochemical techniques has led to a better understanding of the basic mechanisms occurring during tribocorrosion. The scanning vibrating electrode technique (SVET) made it possible to map the current densities, in the friction track and the vicinity of it [5, 6]. In addition, Assi et al. adapted the electrochemical microcell technique to investigate tribocorrosion processes at the microscale [7, 8]. This technique was then used to determine the tribocorrosion behavior of duplex stainless steels [6].

However, tribocorrosion is complex as it involves multiscale interactions between numerous factors: mechanical, (electro-)chemical, physical, and material factors. The chemical-mechanical mechanisms of tribocorrosion are not yet well understood, partly due to the complexity of the chemical, electrochemical, physical, and mechanical processes involved and to the existence of synergistics effects. In this context, the Pearson correlation method (PCM) can be used as a primary step to determine the relationships between variables [9]. These variables include:

- Surface and subsurface characteristics, namely plastic deformation, roughness, crystallographic texture, defects (cracks and voids), phase changes, and recrystallization.
- Electrochemical parameters describing the corrosion behavior of the system, namely corrosion rate, corrosion potential, volume dissolved, and polarization resistance.
- Experimental conditions of the test (tribocorrosion tests for example).

From the PCM, synergistic effects can also be identified by using the correlation coefficients.

Data derived from the PCM make it possible to develop ideal (simplified) physical models by neglecting some variables and/or some synergistic effects that play a minor role. Ideal physical models are easier to investigate using experimental and numerical techniques than real physical systems that are very complex. Therefore, the use of the PCM in corrosion science and engineering opens a large field of investigation in studying the behavior of complex surfaces generated by industrial processes (such as tribocorrosion).

6.2 The Pearson correlation method

Correlation is a technique for investigating the relationship between quantitative, continuous variables. The PCM is one of these technique. In statistics, four types of correlations can be measured, namely Pearson correlation (PCM), Kendall rank correlation, Spearman correlation, and the Point-Biserial correlation. The PCM is the most widely used correlation statistic to measure the degree of the relationship between variables.

6.2.1 Assumptions

There are five assumptions that underpin the PCM. If any of these assumptions are not met, the Pearson's correlation might not lead to a valid result. These assumptions are listed and discussed below:

- **Assumption #1: level of measurement**. Each variable should be continuous. If one or both of the variables are ordinal in measurement, then a Spearman correlation [10] could be conducted instead. Examples of continuous variables include temperature and pH.
- **Assumption #2: related pairs**. Each observation should have a pair of values. If the correlation is between weight and height, then each observation used should have both a weight and a height value.
- **Assumption #3: linearity**. There needs to be a linear relationship between variables. If the relationship displayed in the scatterplot is not linear, data must be "transformed" or the Spearman's correlation [10] must be used instead of the PCM. The Spearman rank correlation is a nonparametric test that is used to measure the degree of association between two variables. The Spearman rank correlation test does not carry any assumptions about the distribution of the data and is the appropriate correlation analysis when the variables are measured on a scale that is at least ordinal.

- **Assumption #4: absence of outliers.** There should be no significant outliers. Outliers are simply single data points within your data that do not follow the usual pattern. Correlations are very sensitive to outliers and a single unusual observation may have a huge impact on a correlation. Such outliers are easily detected by a quick inspection of the scatterplot.
- **Assumption #5: normality of variables.** Variables should be approximately normally distributed. In order to assess the statistical significance of the Pearson correlation, you need to have bivariate normality, but this assumption is difficult to assess, so a simpler method is more commonly used. This is known as the Shapiro-Wilk test of normality.

6.2.2 Basic principles of the PCM

The most common measure of correlation in the PCM is the Pearson's correlation coefficient calculated according to the formula shown in Eq. (6.1). The Pearson's correlation coefficient is defined as the covariance of X and Y (two variables) divided by the product of their standard deviations [11]. In this chapter, these coefficients will be noted $\rho_{X,Y}$.

$$\rho_{X,Y} = \frac{\text{cov}(X,Y)}{\sigma_X \times \sigma_Y} \tag{6.1}$$

where cov(X,Y) is the covariance of (X,Y), σ_X is the standard deviation of X, σ_Y is the standard deviation of Y. The covariance of (X,Y) and the standard deviation of X and Y are defined in Eqs. (6.2)–(6.4).

$$\text{cov}(X,Y) = \frac{\sum_{i=1}^{n}(x_i - \bar{x})(y_i - \bar{y})}{n-1} \tag{6.2}$$

$$\sigma_X = \sqrt{\frac{1}{n}\sum_{i=1}^{n}(x_i - \bar{x})^2} \tag{6.3}$$

$$\sigma_Y = \sqrt{\frac{1}{n}\sum_{i=1}^{n}(y_i - \bar{y})^2} \tag{6.4}$$

where n is the number of observations (X,Y), x_i and y_i are the i^{rd} values of X and Y, respectively, $\bar{x}$ and $\bar{y}$ are the arithmetic mean values of all x_i and y_i, respectively. An estimation of the Pearson's correlation coefficient, $r_{X,Y}$, is given in Eq. (6.5a and b).

$$r_{X,Y} = \frac{\sum_{i=1}^{n}(x_i - \bar{x})(y_i - \bar{y})}{\sqrt{\sum_{i=1}^{n}(x_i - \bar{x})^2 \sum_{i=1}^{n}(y_i - \bar{y})^2}} \tag{6.5a}$$

$$r_{X,Y} = \frac{\sum_{i=1}^{n} x_i y_i - \frac{\sum_{i=1}^{n} x_i \sum_{i=1}^{n} y_i}{n}}{\sqrt{\left(\sum_{i=1}^{n} x_i^2 - \frac{\left(\sum_{i=1}^{n} x_i\right)^2}{n}\right) \times \left(\sum_{i=1}^{n} y_i^2 - \frac{\left(\sum_{i=1}^{n} y_i\right)^2}{n}\right)}} \quad (6.5b)$$

For example, for the given variable set shown in Table 6.1, the Pearson's correlation coefficient is equal to $r_{X,Y} = 0.82$.

6.2.3 Significance test

A significance test was must be carried out to decide whether the correlation between variables is significant. For this purpose, the corresponding t-value is calculated using Eq. (6.6).

$$t = \frac{r_{X,Y}}{\sqrt{\frac{1 - r_{X,Y}^2}{n-2}}} \quad (6.6)$$

Once the t-value is calculated, we have to:

- calculate the degrees of freedom of the system df, $df = n - 2$,
- select a significance level, α, (usually $1 - \alpha = 0.95$ or 0.99),
- determine if we want to use a one-tailed or a two-tailed t-test.

A critical value, t_{crit}, can then be extracted from the t-distribution table (Table 6.2) using these data. The t-value is statistically significant at a given level when it is equal to or greater than t_{crit}. This means:

- If t-value $> t_{crit}$, the null hypothesis (no correlation) is rejected.
- If t-value $< t_{crit}$, the alternative hypothesis (there is correlation) is rejected.

For example, considering the distribution shown in Table 6.1 ($r_{X,Y} = 0.82$ and $n = 11$), one have:

$$t = \frac{0.82}{\sqrt{\frac{1 - 0.82^2}{11 - 2}}} = 4.297 \text{ and } df = 9 \quad (6.7)$$

For a significance level of $\alpha = 0.05$ (with a two-tails t-distribution), the critical value t_{crit} derived from the t-distribution table (Table 6.2) is equal to 2.262. Therefore, $t = 4.297 \gg 2.262$. In this case, $r_{X,Y}$ is significant.

Table 6.1 Example of calculation of $r_{X,Y}$.

i	1	2	3	4	5	6	7	8	9	10	11	Average	σ
x	10	8	13	9	11	14	6	4	12	7	5	9	3.32
y	8.04	6.95	7.58	8.81	8.33	9.96	7.24	4.26	10.84	4.82	5.68	7.501	2.03

Table 6.2 Critical values, t_{crit}, for the Student's t-test (two-tails and $1 \leq df \leq 20$).

df	$\alpha = 0.2$	$\alpha = 0.1$	$\alpha = 0.05$	$\alpha = 0.02$	$\alpha = 0.01$	$\alpha = 0.001$
1	3.078	6.314	12.706	31.821	63.656	636.578
2	1.886	2.92	4.303	6.965	9.925	31.6
3	1.638	2.353	3.182	5.541	5.841	12.924
4	1.533	2.132	2.776	3.747	4.604	8.61
5	1.476	2.015	2.571	3.365	4.032	6.869
6	1.44	1.943	2.447	3.143	3.707	5.959
7	1.45	1.895	2.365	2.998	3.499	5.408
8	1.397	1.86	2.306	2.896	3.355	5.041
9	1.383	1.833	2.262	2.821	3.25	4.781
10	1.372	1.812	2.228	2.764	3.169	4.587
11	1.363	1.796	2.201	2.718	3.106	4.437
12	1.356	1.782	2.179	2/681	3.055	4.318
13	1.35	1.771	2.16	2.65	3.012	4.221
14	1.345	1.761	2.145	2.624	2.977	4.14
15	1.341	1.753	2.131	2.602	2.947	4.073
16	1.337	1.746	2.12	2.583	2.921	4.015
17	1.333	1.74	2.11	2.567	2.898	3.965
18	1.33	1.734	2.101	2.552	2.878	3.922
19	1.328	1.729	2.093	2.539	2.861	3.883
20	1.325	1.725	2.086	2.528	2.845	3.85

Modified from B.L. Weathington, C.J.L. Cunningham, D.J. Pittenger, Understanding Business Research, first ed., John Wiley & Sons, Inc, Appendix B, 2012, pp. 435–483.

Some handbooks also give critical values for the Pearson's correlation coefficient, Table 6.3. Considering the distribution shown in Table 6.1 ($r_{X,Y} = 0.82$ and $n = 11$), a two-tailed distribution and a significance level of $\alpha = 0.05$, the critical value is 0.6021. Therefore, we confirm that $r_{X,Y}$ (=0.82) is significant (greater than the critical value).

6.2.4 Output of the PCM

Most spreadsheet editors (such as Excel, Google sheets, and OpenOffice) can compute correlations. The output of the PCM is a correlation coefficient r (for a pair of variables). It takes values between +1 and −1. The stronger the correlation of the two variables, the closer the coefficient r will be to either +1 or −1. A value of 0 indicates that there is no correlation between the two variables. A value greater than 0 indicates a positive correlation. This means that the variable X increases as the variable Y increases, Fig. 6.1A and B. A value less than 0 indicates a negative correlation. In this case, the variable X decreases as the variable Y increases.

Table 6.3 Critical values for the Pearson's correlation coefficient (two-tails and $1 \leq df \leq 20$).

df	$\alpha = 0.2$	$\alpha = 0.1$	$\alpha = 0.05$	$\alpha = 0.02$	$\alpha = 0.01$	$\alpha = 0.001$
1	0.9511	0.9877	0.9969	0.9995	0.9999	0.9999
2	0.8	0.9	0.95	0.98	0.99	0.999
3	0.687	0.8054	0.8783	0.9343	0.9587	0.9911
4	0.6084	0.7293	0.8114	0.8822	0.9172	0.9741
5	0.5509	0.6694	0.7545	0.8329	0.8745	0.9509
6	0.5067	0.6215	0.7067	0.7887	0.8343	0.9249
7	0.4716	0.5822	0.6664	0.7498	0.7977	0.8983
8	0.4428	0.54494	0.6319	0.7155	0.7646	0.8721
9	0.4187	0.5214	0.6021	0.6851	0.7348	0.847
10	0.3981	0.4973	0.576	0.6581	0.7079	0.8233
11	0.3802	0.4762	0.5529	0.6339	0.6835	0.8010
12	0.3646	0.4575	0.5324	0.612	0.6614	0.78
13	0.3507	0.4409	0.514	0.5923	0.6411	0.7604
14	0.3383	0.4259	0.4973	0.5742	0.6226	0.7419
15	0.3271	0.4124	0.4821	0.5577	0.6055	0.7247
16	0.317	0.4	0.4683	0.5425	0.5897	0.7084
17	0.3077	0.3887	0.4555	0.5285	0.5751	0.6932
18	0.2992	0.3783	0.4438	0.5155	0.5614	0.6788
19	0.2914	0.3687	0.4329	0.5034	0.5487	0.6652
20	0.2841	0.3598	0.4227	0.4921	0.5368	0.6524

Modified from B.L. Weathington, C.J.L. Cunningham, D.J. Pittenger, Understanding Business Research, first ed., John Wiley & Sons, Inc, Appendix B, 2012, pp. 435–483.

If more than two variables are involved as it is often the case in real systems, correlations are usually shown in a square table known as a correlation matrix, Table 6.4. In this case, the diagonal elements are systematically equal to 1 (correlations between each variable and itself).

It has to be kept in mind that the existence of a strong correlation does not imply a causal link between the variables. In addition, the interpretation of a correlation coefficient depends on the context. A correlation of 0.8 may be very high or relatively low. However, the strength of the correlation can be verbally described using the guide that Evans [12] suggests for the absolute value of $r_{X,Y}$:

- $r_{X,Y} = 0.00$–0.19 → "very weak"
- $r_{X,Y} = 0.20$–0.39 → "weak"
- $r_{X,Y} = 0.40$–0.59 → "moderate"
- $r_{X,Y} = 0.60$–0.79 → "strong"
- $r_{X,Y} = 0.80$–1.00 → "very strong"

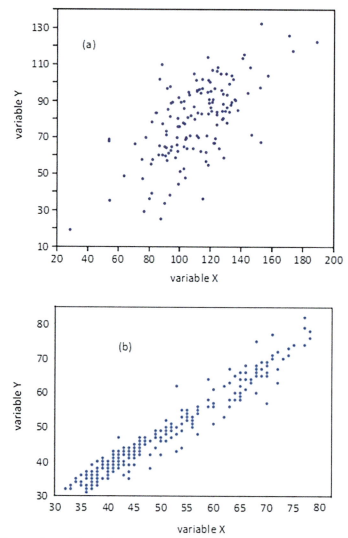

Fig. 6.1 Scatter plots of Y as a function of X with (A) $r = 0.67$ and (B) $r = 0.97$.

Other important remarks about the Pearson's correlation coefficients, r, are listed:

- r is expressed without units. It does not change if we change the units or the origin of X or Y (invariant coefficients),
- r is the symmetric: $r_{X,Y} = r_{Y,X}$,
- The correlation coefficient is very sensitive to extreme points, Fig. 6.2A, and outliers (see assumptions #4),
- The correlation coefficient is very sensitive to the distributions of points, Fig. 6.2B.

The use of the Pearson's correlation coefficients

Table 6.4 Example of Pearson correlation matrix with 11 variables.

	A	B	C	D	E	F	G	H	I	J	K
A	1										
B	0.02	1									
C	0.05	**0.3**	1				Symetrical				
D	**0.55**	0.04	0.15	1							
E	−0.13	**0.36**	0.05	−0.13	1						
F	**0.31**	**0.18**	0.02	**0.46**	−0.02	1					
G	**0.27**	0.11	0.15	**0.39**	**0.26**	**0.22**	1				
H	**0.51**	0.02	0.12	**0.84**	−0.04	**0.47**	**0.5**	1			
I	**0.3**	**0.19**	0.03	**0.27**	**0.28**	0.05	0.11	**0.21**	1		
J	**0.25**	**0.44**	0.14	**0.49**	0.03	**0.41**	**0.31**	**0.54**	**0.22**	1	
K	**0.51**	0.1	0.1	**0.67**	−0.2	**0.36**	**0.3**	**0.63**	**0.23**	**0.3**	1

Numerical values in bold print are significant ($P < .05$ in this case).

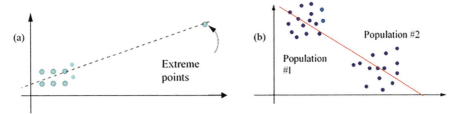

Fig. 6.2 Examples of distribution of points in a population.

6.3 Methodology

The following methodology may be adopted to perform statistical analysis of results derived from tests (tribocorrosion tests for example). It is divided into three main steps:

- **Step 1**. The experimental conditions of the experimental tests are defined by applying the theory of the design of experiments [13]. One of the objectives of this design is to optimize the number of tests. While the tests are running, some parameters may be recorded (potential, current, pH, etc. during tribocorrosion tests, friction coefficient, normal load, for example). This can be performed at the local scale (using miroprobes) or the global scale.
- **Step 2**. All surface and subsurface characteristics are evaluated using different experimental methods. This includes mechanical, metallurgical, and physical properties. Note that numerical simulation may be helpful to evaluate some characteristics that cannot be determined experimentally. Surface and subsurface characteristics may be evaluated prior to tests and after tests. The weight loss can also be measured.
- **Step 3**. Statistical analysis based on the PCM is used to evaluate relationships between experimental conditions (tribocorrosion tests) vs surface and subsurface characteristics (surface integrity) vs corrosion parameters.

6.4 Application of the PCM in corrosion science and engineering

The PCM was used to solve important issues in corrosion science and engineering. It provided important information in various industrial sectors, such as construction and civil engineering [14–16], medicine [17], manufacturing industries [18, 19], microbiologically influenced corrosion [20], energy [21], and others [9]. It also provided crucial information to better understand basic mechanisms.

The PCM was applied to optimize the design of corrosion protection for steel structures using an organic coating (using epoxies) [14]. A key issue is whether a higher coating thickness over cut edges or a better edge coverage, improves the corrosion protection performance of epoxy-based coatings. The PCM provided statistically funded arguments for this important issue. It was shown that no linear correlation exists between edge coverage and corrosion protection performance over cut edges. Therefore, the authors concluded that good edge coverage is not a sufficient preposition for a good corrosion protection performance underwater ballast tank conditions.

Corrugated steel plate culverts may be exposed to the risk of soil corrosion. The PCM helped to understand the interactions between key parameters involved (pH vs the backfill resistivity, and moisture content vs the backfill resistivity) [15], Table 6.5. A strong correlation between the backfill resistivity and pH during summer and spring, in addition to a correlation between resistivity and moisture content during the spring season was found. Statistical analysis indicates that resistivity decreases when pH increases (negative correlation coefficients, Table 6.5); whereas, resistivity increases with increasing moisture content (positive correlation coefficient).

Such statistical results make it possible to better understand basic mechanisms of soil corrosion of culverts and to select appropriate sites where culverts can be installed with a low risk of short-term corrosion.

The PCM was also employed to identify the relationships between the concentration of some species (lipopolysaccharide (LPS) and dextrose (DEX)) in the simulated body fluid (SBF) and the corrosion behavior of Ti-6Al-4V alloy after different surface preparations [17]. Optimizing the surface preparation of implants is a key issue for increasing the lifetime of implants. Two different surface preparations were considered: mechanical polishing (smooth surfaces) and etching by using nitric, sulfuric, and hydrochloric acids (treated surfaces). In the statistical analysis, correlations between corrosion parameters (corrosion potential E_{corr}, corrosion current density I_{corr}, the

Table 6.5 Pearson's correlation coefficients for the analyzed culverts.

	Correlation coefficients	
	Summer	Spring
pH vs backfill resistivity	−0.6762	−0.9395
Moisture content vs backfill resistivity	0.3288	0.8390

Modified from D. Beben, J. Perform. Constr. Facil. 29 (2015).

current density in the passive range I_{pass}, double-layer capacitance C_{dl}, and polarization resistance R_p) and concentrations of species used alone (noted Dextrose or LSP in Table 6.6) and combined (groups noted Dextrose + LPS in Table 6.6) were investigated. Table 6.6 shows numerical values of the correlation coefficients derived from the PCM for both smooth and treated surfaces.

In Table 6.6, the parameter r is the Pearson's correlation coefficient. The parameter p (or P-value) is the probability of obtaining a test statistic at least as extreme as the one that was actually observed, assuming that the null hypothesis (no correlation, see Section 6.2.3) is true. It can be easily calculated by using many commercial software packages. The null hypothesis is rejected when the P-value turns out to be less than a certain significance level, α. This significance level is arbitrarily chosen ($\alpha = 0.05$ in Table 6.6).

From Table 6.6, it can be seen that there is a correlation only between the concentration of dextrose + lipopolysaccharide and the corrosion potential (E_{corr}). The corrosion potential decreases as the concentration increases (negative coefficient). For all the other configurations, no correlations were found. The concentration of species does not affect significantly I_{pass}, I_{corr}, C_{dl}, and R_p.

Table 6.6 Numerical values of the correlation coefficients, r, derived from the PCM for both the smooth and treated surfaces.

	\multicolumn{6}{c}{Smooth surfaces}					
	\multicolumn{2}{c}{DEX}	\multicolumn{2}{c}{LPS}	\multicolumn{2}{c}{DEX + LSP}
	r	p	r	p	r	p
E_{corr}	−0.479	0.192	−0.615	0.078	**−0.589**[a]	**0.001**
I_{corr}	−0.664	0.051	0.469	0.203	0.250	0.208
i_{pass}	0.534	0.139	0.069	0.806	−0.134	0.504
C_{dl}	0.070	0.858	0.477	0.194	0.008	0.967
R_p	−0.557	0.119	0.224	0.562	0.217	0.277
	\multicolumn{6}{c}{Treated surfaces}					
	\multicolumn{2}{c}{DEX}	\multicolumn{2}{c}{LPS}	\multicolumn{2}{c}{DEX + LSP}
	r	p	r	p	r	p
E_{corr}	0.639	0.064	−0.246	0.523	−0.121	0.547
I_{corr}	0.382	0.311	0.347	0.361	**0.391**[a]	**0.044**
i_{pass}	0.500	0.170	**0.789**[a]	**0.011**	**0.435**[a]	**0.023**
C_{dl}	−0.172	0.658	**0.808**[a]	**0.008**	0.104	0.605
R_p	**0.960**[a]	**0.0001**	**−0.931**[a]	**0.0001**	0.259	0.193

Bold values indicate significant correlation. *DEX*, dextrose; *LPS*, lipopolysaccharide.
[a] Significant at the 0.05 level (independent *t*-test).

Modified from L.P. Faverani, W.G. Assuncao, P.S.P. de Carvalho, J.C.C. Yuan, C. Sukotjo, M.T. Mathew, V.A. Barao, PLoS One 9 (2014) e93377.

After etching (treated surfaces in Table 6.6), there is a strong correlation between LSP concentration and current density in the passive range (I_{pass}), double-layer capacitance (C_{dl}), and polarization resistance (R_p) values. On the other hand, the combination of dextrose and lipopolysaccharide (LSP) was correlated with the corrosion current density (I_{corr}) and current density in the passive range (I_{pass}).

Machining is widely used by nearly all manufacturing industries. However, selecting wrong machining conditions (values of the cutting speed, feed, depth of cut…) may increase drastically the corrosion susceptibility of machined workpieces. Under this context, the PCM was used to determine the influence of surface and subsurface characteristics on the corrosion behavior of metals and alloys after machining [18, 19]. In these studies, surface and subsurface characteristics include quadratic stress, crystallographic texture, lubrication, and surface roughness. All these characteristics were first quantified after machining using various techniques (surface profilometer, X-ray diffraction coupled with the $\sin^2 \psi$ method and X-ray diffraction using a texture goniometer). Different corrosion tests were considered:

- **Immersion tests in aqueous solution at the OCP value**. The variable describing the corrosion behavior used in the PCM was the OCP. It was measured locally, in grooves and ridges generated by machining, by means of the electrochemical microcell technique [18]. The evolution of the OCP is an important parameter describing the electrochemical behavior of samples under free corrosion. Fig. 6.3 shows the values of the correlation coefficients between surface/subsurface characteristics and the OCP values (in ridges and grooves) and synergistic effects derived from the PCM (noted by "interactions" in Fig. 6.3). It can be mentioned that similar trends were obtained in grooves and ridges. From the PCM, the authors found that the OCP was shifted in the anodic direction with increasing quadratic stress, roughness and with the use of a lubricant (classified according to their increasing influence on the OCP).
- **Salt fog atmosphere tests**. Under these conditions [19], two compounds are formed on copper after machining: atacamite/paratacamite and a black layer consisting of a mixture of residues of the lubricant used during machining and which are deposited on the surface during machining and species from the salt atmosphere. The authors found that the formation of atacamite/paratacamite was associated with severe degradation of the samples. Statistical analysis based on the PCM revealed that only surface roughness and quadratic stress have a significant influence on the formation of atacamite/paratacamite (and therefore on the degradation process). By contrast, the formation of the black layer was controlled by lubrication and the crystallographic texture. It is interesting to note that in this work, the evolution of the correlation coefficients was quantified vs. time using interrupted experiments. It was then possible to identify relationships between variables and synergistic effects vs time. This statistical analysis is then resolved over time.

Therefore, the PCM can be used to optimize conditions selected for processing operations (such as machining) to reduce deleterious effects of corrosion and to increase the lifetime of engineering systems.

Heat-exchange tubes of vapor condensers may undergo pitting corrosion in power plants [21]. This is primarily connected with the cooling water composition and temperature. A satisfactory statistical description of the corrosion defect (pits) depth on the cooling water size was provided by employing the Weibull distribution. The PCM was used as a criterion for the correlation of the theoretical distribution of experimental data. The correlation coefficient is greater than 0.99, indicating an excellent description of the defect depth distribution law using the Weibull function. Determining and validating distribution laws are of major importance in predicting the lifetime of condensers.

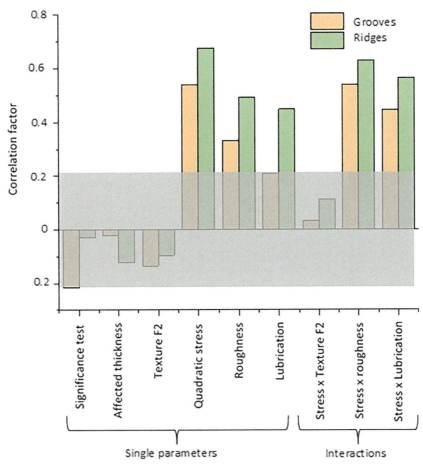

Fig. 6.3 Histogram graph derived from the Pearson's correlation matrix.
Modified from J. Gravier, V. Vignal, S. Bissey-Breton, J. Farre, Corros. Sci. 50 (2008) 2885–2894.

6.5 Application of the PCM to stress corrosion cracking and fatigue corrosion

In stress corrosion cracking (SCC) and fatigue corrosion (FC), the transition from a corrosion pit to a crack is a crucial issue. Understanding this transition is of major importance to predict the lifetime of engineering systems. Correlation analysis based on the PCM was performed to discern the best correlated parameters. In these investigations, dog-bone specimens (with no defects) were used. In addition, it is assumed that a single fatigue crack initiates and propagates. Under these conditions, correlated parameters were found to depend on the alloy considered:

- pit depth × aspect ratio (representing the synergistic effect between pit depth and aspect ratio) for D6ac steel [22] and AA7050 aluminum alloy [23],
- pit depth for AA7075-T6511 [24],
- pit cross-sectional area for AA7075-T6 [25] and AA7010-T7651 [26].

In the case of tensile specimens of AA7075-T6, Huang et al. [27] showed that correlated parameters depend on the presence of defects (for example, a central hole modelling the effect of precorrosion) and on the mode of propagation (single-crack initiations or multicrack initiations):

- for single-crack initiations, (equivalent crack depth/aspect ratio) vs (critical pit depth/aspect ratio) are best correlated with correlation coefficient of 0.9,
- for multicrack initiations, the best correlated parameters are (equivalent crack depth/aspect ratio) vs (aspect ratio) with correlation coefficient of 0.69.

From this statistical analysis, equivalent crack size models were developed [27] with these best-correlated parameters for single-crack and multicrack initiations, respectively. The authors underlined that the precorrosion fatigue lives predicted with these models agree well with the experimental results.

6.6 Application of the PCM to tribocorrosion

The PCM was used in many works to investigate the tribological behavior of metallic alloys and workpieces [28–33]. By contrast, only a few papers (recently published) were devoted to solving issues in tribocorrosion [34–37].

A few years ago, novel bioactive Ti coatings doped with various elements (Ca, P, Si, and Ag) were elaborated by microarc oxidation (MAO) [35]. Aqueous mixtures used during MAO were prepared from $Ca(CH_3CO_2)_2$, $C_3H_7Na_2O_6P$, Na_2SiO_3, and Ag, Table 6.7. Prior to MAO, mechanical polishing was applied to the samples to achieve mirror finishing. Some samples were also sandblasted and acid-etched (Al-oxide).

The stability of samples was then tested underwear and corrosion during tribocorrosion experiments. Fig. 6.4A shows the numerical values of the Vicker's microhardness and total mass loss (degradation due to tribocorrosion processes) measured for the different MAO conditions. Correlation analysis between these two parameters was then examined using the PCM with $\alpha \leq 0.05$ (significance level, see Section 6.2.3), Fig. 6.4B.

The PCM showed that there is a strong correlation between the Vicker's microhardness of the coating and total mass loss measured after tribocorrosion tests, Fig. 6.4B. Negative values of the correlation coefficient indicated that the total mass loss increases as the microhardness decreases. These results contribute to a better understanding of tribocorrosion wear mechanisms. In addition, the correlations found using the PCM then make it possible to predict the evolution of the mass loss without carrying out tribocorrosion tests, by simply carrying out microhardness measurements. These measurements are easy and fast.

Table 6.7 Experimental details of the MAO treatments.

	1-CaP5	1-CaP10	1-CaPAg5	1-CaPAg10	2-CaP5	2-CaP10	2-CaPSi5	2-CaPSi10
Ca(CH$_3$CO$_2$)$_2$	0.3M				0.1M			
C$_3$H$_7$Na$_2$O$_6$P	0.02M				0.03M			
Na$_2$SiO$_3$							0.04 M	
Ag			0.62 g/L					
Duration	5	10	5	10	5	10	5	10

Modified from I.S. Marques, M.F. Alfaro, M.T. Saito, N.C. da Cruz, C. Takoudis, R. Landers, M.F. Mesquita, F.H. Nociti Junior, M.T. Mathew, C. Sukotjo, V.A. Barão, Biointerphases 11 (2016) 031008.

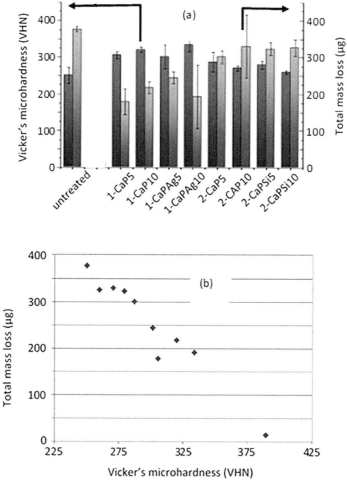

Fig. 6.4 (A) Numerical values of the Vicker's microhardness and total mass loss vs MAO conditions (see Table 6.4) and (B) Correlation between Vicker's microhardness and total mass loss.
Modified from I.S. Marques, M.F. Alfaro, M.T. Saito, N.C. da Cruz, C. Takoudis, R. Landers, M.F. Mesquita, F.H. Nociti Junior, M.T. Mathew, C. Sukotjo, V.A. Barão, Biointerphases 11 (2016) 031008.

The tribocorrosion processes have their origins at the nanoscale. Therefore, wear measurements at the nanoscale is a crucial step to understand the relationships between wear processes, the alloy microstructure (influence of grain size and presence of carbides), and the surface roughness. Different techniques can be applied to perform such measurements, namely atomic force microscopy (AFM), scanning electron microscopy (SEM), and transmission electron microscopy (TEM). A description of the principles, advantages, and limitations of these techniques in this field of tribology has been published [38]. They have been applied to study the nanotribological and nanomechanical properties of materials [39]. Recently [36, 37], a new experimental methodology is being established to give a better understanding of the deformation-corrosion processes occurring at the nanoscale during tribocorrosion. This work aims to understand the relationships between wear processes at the nanoscale and the microstructure of CoCrMo alloys (for loads in the range of 5–30 mN). In a first step, a new white-light interferometry system (obtaining 3-D data maps and section profiles) was used to measure nanoscale wear damage. Obtained results were compared to those derived from conventional measuring systems (AFM and nanoTest Vantage system). The PCM was then used to validate depth measurements at the nanoscale with these different techniques, Fig. 6.5.

It was concluded (Fig. 6.5) that for the highest loads (i.e., >27 mN in Fig. 6.5C), the AFM triples the values measured by the NanoTest Vantage (nanoindenter system) and the interferometer systems. Also the interferometer values show a bigger dispersion at some points compared with the AFM or the NanoTest Vantage system.

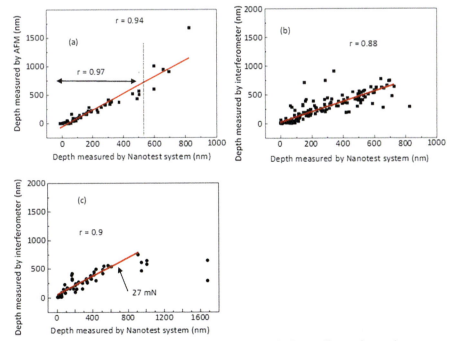

Fig. 6.5 (A–C) Comparison of depth measurements on CoCrMo alloys using various experimental techniques (*r* = Pearson correlation coefficient).
Modified from V. Martinez Nogues, Nanoscale Tribo-Corrosion of CoCrMo Biomedical Alloys (thesis), University of Southampton, Faculty of Engineering and the Environment, 2016; V. Martinez-Nogues, J.M. Nesbitt, R.J.K. Wood, R.B. Cook, Tribol. Int. 93 (Part B) (2016) 563–572.

6.7 Conclusions

PCM is the test statistics that measures the statistical relationship between two continuous variables. It is known as a popular method of measuring the relationship between variables of interest because it is based on the method of covariance. It gives useful information about the magnitude of the relationship, or correlation, as well as the direction of the relationship.

The PCM was frequently used in corrosion science and engineering (mainly pitting corrosion, stress corrosion cracking, and fatigue corrosion) with the aim of:

- optimizing design of corrosion protection,
- optimizing industrial processes (such as machining) to reduce the corrosion susceptibility of manufactured workpieces on the field,
- developing simplified physical models representing the real systems (that are very complex),
- validating novel experimental methodologies.

The PCM can be used to understand basic mechanisms by determining the influence of surface and subsurface parameters and the role of synergistic effects. These effects often play a major role in corrosion processes.

The PCM can be used to investigate the tribological properties of alloys. By contrast, the PCM was rarely employed till today in tribocorrosion research. However, as it is the case in pitting corrosion, SCC and FC, this statistical method opens novel fields of investigation in understanding, testing, standardization, and application to engineering practice.

6.8 Notations

$X = (x_1, x_2, x_3, \ldots, x_i, \ldots, x_n)$ variables composed of n data
$Y = (y_1, y_2, y_3, \ldots, y_i, \ldots, y_n)$ variables composed of n data
x_i i^{rd} values of X
y_i i^{rd} values of Y
n number of observations (X,Y)
$cov(X,Y)$ covariance of (X,Y), units are those of X times those of Y
σ_X standard deviation of X, same unit as X
$\bar{x}$ arithmetic mean value of X, same unit as X
$\rho_{X,Y}$ Pearson's correlation coefficient
$r_{X,Y}$ approximation of the Pearson's correlation coefficient
t t-value
t_{crit} critical value of t-value derived from t-distribution table
α significance level
df degree of freedom
E_{corr} corrosion potential (V)
i_{corr} corrosion current density (A m^{-2})
i_{pass} passivation current density (A m^{-2})
C_{dl} double-layer capacitance (F m^2)
R_p polarization resistance (ohm m^2)

References

[1] D. Landolt, S. Mischler, M. Stemp, Electrochim. Acta 46 (2001) 3913–3929.
[2] P. Ponthiaux, F. Wenger, D. Drees, J.P. Celis, Wear 256 (2004) 459–468.
[3] M. Keddam, F. Wenger, Electrochemical methods in tribocorrosion, in: D. Landolt, S. Mischler (Eds.), Tribocorrosion of Passive Metals and Coatings, Part II, Woodhead Publishing Series in Metals and Surface Engineering, 2011, pp. 187–221.
[4] S. Mischler, Tribol. Int. 41 (2008) 573–583.
[5] H. Krawiec, V. Vignal, P. Ponthiaux, F. Wenger, ECST 3 (2007) 355–361.
[6] H. Krawiec, V. Vignal, O. Heintz, P. Ponthiaux, F. Wenger, J. Electrochem. Soc. 155 (2008) C127–C130.
[7] F. Assi, T. Suter, H. Böhni, Lubr. Sci. 6 (2006) 17–28.
[8] F. Assi, H. Böhni, Wear 233–235 (1999) 505–514.
[9] S.K. Kolawole, F. Kolawole, O.P. Enegela, O. Adewoye, Adv. Mat. Res. 1132 (2015) 349–365.
[10] J.H. Zar, Spearman rank correlation, in: P. Armitage, T. Colton (Eds.), Encyclopedia of Biostatistics, second ed., vol. VII, Wiley, Hoboken, NJ, 2005, pp. 5095–5101.
[11] K. Pearson, Philos. Mag. 50 (1900) 157–175.
[12] J.D. Evans, Straightforward Statistics for the Behavioral Sciences, Brooks/Cole Publishing, Pacific Grove, CA, 1996.
[13] G. Taguchi, S. Konishi, Orthogonal Arrays and Linear Graph, American Supplier Institute press, 1987, pp. 35–38.
[14] A.W. Momber, S. Buchbach, P. Plagemann, T. Marquardt, Prog. Org. Coat. 108 (2017) 90–92.
[15] D. Beben, J. Perform. Constr. Facil. 29 (2015).
[16] A.N. Grishina, A.N. Zemlyakov, E.V. Korolev, K.Y. Okhotnikova, V.A. Smirnov, Identification of the corrosion in cement composites by means of statistical modelling, in: Vestnik MGSU [Proceedings of Moscow State University of Civil Engineering], 2014, pp. 87–97. no. 4.
[17] L.P. Faverani, W.G. Assuncao, P.S.P. de Carvalho, J.C.C. Yuan, C. Sukotjo, M.T. Mathew, V.A. Barao, PLoS One 9 (2014), e93377.
[18] J. Gravier, V. Vignal, S. Bissey-Breton, J. Farre, Corros. Sci. 50 (2008) 2885–2894.
[19] J. Gravier, V. Vignal, S. Bissey-Breton, Corros. Sci. 61 (2012) 162–170.
[20] K.E. Duncan, B.M. Perez-Ibarra, G. Jenneman, J.B. Harris, R. Webb, K. Sublette, Appl. Microbiol. Biotechnol. 98 (2014) 907–918.
[21] V.A. Golovin, N.V. Pechnikovb, V.A. Shchelkova, A.Y. Tsivadze, Prot. Met. Phys. Chem. Surf. 54 (2018) 1221–1232.
[22] T. Mills, P.K. Sharp, C. Loader, The incorporation of pitting corrosion damage into F-111 fatigue life modelling, in: DSTO-RR-0237, Fishermans Bend Vic 3207, Australia, 2002.
[23] P.K. Sharp, S.A. Barter, G. Clark, Localised life extension specification for the F/A-18 Y470 X-19 pocket, in: DSTO-TN-0279, Fishermans Bend Vic 3207, 2000.
[24] M. Liao, G. Renaud, N.C. Bellinger, Int. J. Fatigue 29 (2007) 677–686.
[25] A. Shekhter, C. Loader, W. Hu, B.R. Crawford, Assessment of the effect of pitting corrosion on the safe life prediction of the P-3C, in: DSTO-TR-2080, Fishermans Bend Vic 3207, Australia, 2007.
[26] B.R. Crawford, C. Loader, A.R. Ward, C. Urbani, M.R. Bache, S.H. Spence, D.G. Hay, W. J. Evans, G. Clark, A.J. Stnham, Fatigue Fract. Eng. Mater. Struct. 28 (2005) 795–808.
[27] Y. Huang, X. Ye, B. Hu, L. Chen, Int. J. Fatigue 88 (2016) 217–226.

[28] N.F.M. Yusof, Z.M. Ripin, Tribol. Trans. 57 (4) (2014) 715–729.
[29] S.A. Adnani, S.J. Hashemi, A. Shooshtari, M.M. Attar, Tribol. Ind. 35 (2013) 61–68.
[30] L.A. Dobrzañski, M. Kremzer, J. Trzaska, A. Wodarczyk-Fligier, Arch. Mater. Sci. Eng. 30 (2008) 37–40.
[31] M. Kalin, M. Polajnar, Tribol. Lett. 52 (2013) 185–194.
[32] Z. Geng, D. Puhan, T. Reddyhoff, Tribol. Int. 134 (2019) 394–407.
[33] A. Khanam, B. Mordina, R.K. Tiwari, J. Compos. Mater. 49 (2016) 2497–2507.
[34] S. Nambu, R. Obert, M. Roark, E. Delvechhio, D. Linton, S. Bible, J. Moseley, Accelerated fretting corrosion testing of modular necks for total hip arthroplasty, in: A.S. Greenwald, S.M. Kurtz, J.E. Lemons, W.M. Mihalko (Eds.), Modularity and Tapers in Total Joint Replacement Devices, American Society for Testing and Materials (ASTM) Selected Technical Papers, 1591, 2015, pp. 237–258.
[35] I.S. Marques, M.F. Alfaro, M.T. Saito, N.C. da Cruz, C. Takoudis, R. Landers, M.F. Mesquita, F.H. Nociti Junior, M.T. Mathew, C. Sukotjo, V.A. Barão, Biointerphases 11 (2016), 031008.
[36] V. Martinez Nogues, Nanoscale Tribo-Corrosion of CoCrMo Biomedical Alloys (thesis), University of Southampton, Faculty of Engineering and the Environment, 2016.
[37] V. Martinez-Nogues, J.M. Nesbitt, R.J.K. Wood, R.B. Cook, Tribol. Int. 93 (Part B) (2016) 563–572.
[38] L. Qian, L. Chen, L. Jiang, Wear measurement, in: G.E. Totten (Ed.), Friction, Lubrication, and Wear Technology, American Society for Testing and Materials (ASTM) Handbook, vol. 18, 2017, pp. 225–232.
[39] W. Lu, K. Komvopoulos, J. Tribol. 123 (2001) 641–650.

Fretting wear analysis through a mechanical friction energy approach: Impact of contact loadings and ambient conditions

7

Siegfried Fouvry
Ecole Centrale de Lyon, Université de Lyon, CNRS, UMR 5513, Laboratoire de Tribologie et Dynamique des Systèmes, Ecully, France

7.1 Introduction

In general, fretting occurs between two *tightly* fitting surfaces that are subjected to a cyclic relative motion of extremely small amplitude [1, 2]. For oxidable metals, one of the immediate consequences of the process in normal atmospheric conditions is the production of oxide wear debris, hence the term "fretting corrosion" was previously considered although the actual term of "fretting wear" is more consistent. Fretting movement is usually the result of external vibration, but in many cases, it is the consequence of one of the members of the contact being subjected to cyclic stress (i.e., fatigue), which gives rise to another and usually a more damaging aspect of "fretting fatigue" or "contact fatigue" [2]. Fatigue cracks are nucleated by the fretting contact stresses and propagate due to the bulk fatigue stress. One illustration of fretting wear damage concerns the dovetail interface between the disk and the blades (Fig. 7.1) [3]: During the flight sequence (i.e., takeoff and landing) macro-gross slip slidings are combined with aerodynamical micropartial slidings which can generate combined fretting wear (i.e., wear volume induced by debris formation and debris ejection) but also fretting fatigue (i.e., crack nucleation and crack propagation). Fortunately, new surface treatments combining shot and laser peening joined with thick self-lubricant plasma coating are now applied to prevent any critical damage process. The objective of this chapter is to focus on the fretting wear related to debris formation and ejection so the specific problem of "fretting fatigue" will not be detailed presently.

Hence, there is a crucial interest to formalize how fretting damage occurs, how contact loading conditions like sliding amplitude or contact pressure can monitor the sliding condition and the surface damage evolution, and finally how the fretting wear (wear volume but also wear depth) can be simulated. Finally, some aspects regarding how the ambient condition could influence the fretting wear response will be examined. But before going any further on how loading conditions can influence fretting damage evolution, it is important to better describe the various test configurations usually applied to investigate this aspect.

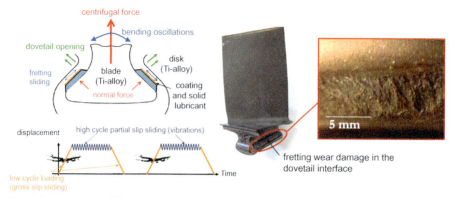

Fig. 7.1 Illustration of fretting wear damage process in turbine engine dovetail interface [4].

7.2 Fretting test experiments

Plain fretting test (i.e., without external bulk stress as applied in fretting-fatigue experiments) is the most usual strategy to investigate fretting wear phenomenon (Fig. 7.2) [5–7]. This test configuration consists of applying a normal force P, a cyclic displacement δ_m (amplitude δ_m^*) using for instance a hydraulic or electromagnetic actuator. The latter alternated displacement results in a tangential force Q (amplitude Q^*). The measured displacement δ_m is obtained from an external extensometer (LVDT, laser) where the loading forces (P and Q) are recorded from load cells.

It is interesting to note that the measured displacement amplitude δ_m^* is not equal to the contact displacement δ^* amplitude [7] (Fig. 7.2). Indeed, under small fretting oscillations, a significant part of the measured displacement is in fact accommodated by the

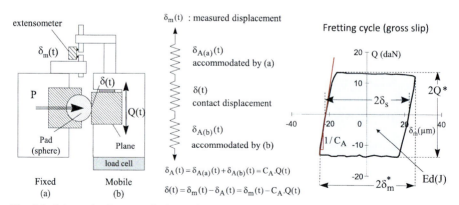

Fig. 7.2 Schematic diagram of a basic fretting wear test and the related fretting cycle. Based on S. Fouvry, P. Kapsa, V. Vincent, Quantification of fretting damage, Wear 200 (1996) (1–2) 186–205.

test assembly deformation (δ_A). This latter depends on the applied tangential force and the compliance of the test assembly (C_A):

$$\delta_m^* = \delta^* + \delta_A = \delta^* + C_A \times Q^* \tag{7.1}$$

The determination of C_A is long and fastidious and requires the stiffness estimation of the test apparatus. One alternative to estimate the real contact sliding amplitude consists in estimating the residual sliding amplitude when $Q = 0$ on the ($\delta_m - Q$) fretting loop such that when $Q = 0$, there is no more tangential deformation of the test apparatus, the measured displacement is then equal to the contact one and therefore:

$$\delta_s = \delta_m(Q=0) = \delta(Q=0) \tag{7.2}$$

Finally, by integrating the fretting loop, the friction energy Ed (J) (i.e., friction work) inputted in the interface can be estimated. Both Ed (J) and δ_s variables can be considered to quantify the wear rate.

7.2.1 Fretting sliding condition

Most of the research work and industrial feedback emphasize that fretting damage chiefly depends on the displacement amplitude and the related sliding condition imposed on the interface (Fig. 7.3) [8, 9].

Partial slip condition: If the displacement amplitude δ remains smaller than a threshold value referred to as sliding transition (δ_t), a partial slip-sliding condition operates. This sliding condition implies a composite contact structure with an inner undamaged stick domain surrounded by a sliding annulus (Fig. 7.3) [8]. Cyclic stress may lead to fatigue crack nucleation and propagation. On the other hand, the friction dissipation (area of the fretting loop) is very low and nearly no surface wear is generated.

Gross slip condition: Above the sliding transition, a full sliding condition operates. The tangential force amplitude reaches a constant value (i.e., $Q^* = \mu \times P$); whereas, the ($\delta_m^* - Q$) fretting hysteresis loop displays a large dissipative open quadratic loop shape. Stabilized gross slip conditions not only favor surface wear by debris formation and debris ejection but also reduce the cracking risk by removing the incipient cracks nucleated on the top surface and by reducing the contact pressure profile.

Reciprocating condition: When the sliding amplitude (δ_s) becomes larger than the contact radius (a), the whole interface will be exposed to the ambient air and the reciprocating sliding condition prevails. The determination of the sliding condition from the ($\delta_m - Q$) fretting loop shape is quite subjective. Hence, applying Mindlin's theory [8] for Hertzian sphere-on-flat contact configuration, various sliding criteria were established allowing a quantitative analysis of the sliding condition (Fig. 7.4) [7, 10, 11].

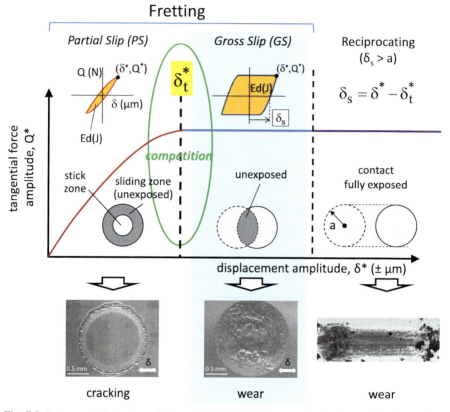

Fig. 7.3 Schematic illustration of the partial-slip fretting, gross-slip fretting, and reciprocation sliding conditions as a function of the displacement amplitude (e.g., sphere-on-flat contact).

The most useful criterion is the energy criterion A expressed as:

$$A = \frac{Ed}{4 \times Q^* \times \delta^*} \tag{7.3}$$

Based on Mindlin's theory, it was demonstrated that there is a constant $A_t = 0.2$ for sphere-on-flat contact below which the contact operates under partial slip (i.e., $A < A_t$) and above which the contact runs under gross slip (i.e., $A > A_t$). Under gross slip condition two friction coefficients can be considered: the nominal coefficient of friction (μ) which is a function of the maximum tangential force that better describes the cracking phenomenon and the so-called energy friction coefficient (μ_e) which provides an averaged description of the friction work during the sliding sequence that better describes the wear processes:

$$\mu = \frac{Q^*}{P} \tag{7.4}$$

Fretting wear

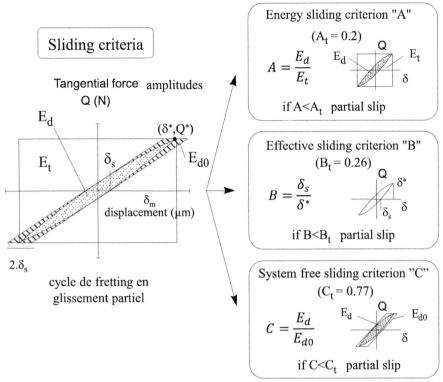

Fig. 7.4 Quantitative analysis of the sliding condition of a Hertzian sphere-on-flat contact using A, B, and C sliding criteria where the sliding transition is related to constants independent of the material properties and contact size: $A_t = 0.2$, $B_t = 0.26$, and $C_t = 0.77$, respectively [7, 10].

$$\mu_e = \frac{Ed}{4 \times P \times \delta_S} \quad (7.5)$$

7.2.2 Fretting regime concept

Most of the industrial interfaces undergo varying loading conditions (Fig. 7.5). Besides, the friction coefficient can evolve during the test and the contact can shift from partial to gross slip conditions.

Therefore, the fretting regime concept was introduced to better describe the stress history imposed on the interface [12, 13]. Stable gross slip conditions are related to the gross slip regime (GSR) promoting surface wear. Constant partial slip condition is denoted by partial slip regime (PSR) which favors the cracking phenomenon. Finally, the mixed fretting regime (MFR), usually observed for intermediate displacement amplitudes close to the sliding transition, promotes competition between wear and cracking phenomena.

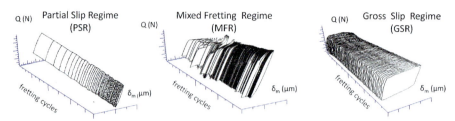

Fig. 7.5 Illustration of fretting regime concept through the fretting log (i.e., plotting of the fretting cycle as a function of the log scale of the fretting cycles).

The fretting regime can be quantified by estimating the proportion of gross slip cycle during the whole test duration defining the GS% parameter so that [14]:

$$GS\% = \frac{N(A > A_t)}{N_{TOTAL}} \qquad (7.6)$$

Stating that:

Partial slip regime (PSR): GS% = 0%
Mixed fretting regime (MFR): 0% < GS% < 100%
Gross slip regime (GSR): GS% = 100%

Hence, using this approach, the competition between fretting wear and fretting cracking as a function of the applied displacement is better rationalized.

Fig. 7.6 compares the evolution of the crack length and wear volume as a function of the applied displacement amplitude δ_m^* and the related GS% fretting regime parameter. As previously underlined, the crack extension starts in PSR, reaches a maximum

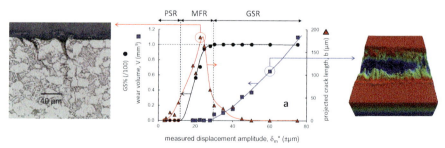

Fig. 7.6 Fretting wear damage chart as a function of the applied displacement amplitude for a cylinder-on-flat contact. Quantification of the wear volume and crack length extension as function of the fretting regime parameter %GS.
Based on S. Heredia, S. Fouvry, Introduction of a new sliding regime criterion to quantify partial, mixed and gross slip fretting regimes: correlation with wear and cracking processes, Wear 269 (2010) (7–8) 515–524.

value in MFR when GS% = 70% then decreases until no crack nucleation takes place under GSR when $\delta_m^* = \pm 50$ µm. The wear volume is negligible under PSR, very small under MSR and increases on a straight-line basis when GSR regime operates (GS% = 100%).

7.3 Fretting mapping approach

The former analysis can be extended by varying the normal force leading to the fretting map approach (i.e., sliding regimes and the related damage reported as a function of the applied displacement amplitude δ_m^* and the applied normal force P for a given contact configuration and a given number of fretting cycles). The Running Condition Fretting Map (RCFM) displays the different PSR, MFR, and GSR regimes [13]. The boundaries are usually characterized by power-law evolutions although various authors consider linear approximations. In fact, assuming Hertzian-Mindlin sphere-on-flat configuration, the transition from partial to gross slip can be expressed as [7]:

$$\delta_t^* = \frac{\mu \times P^{2/3} \times H_M}{R^{1/3}} \tag{7.7}$$

where H_M is an elastic constant.

$$H_M = \left(\frac{3}{4 \times E}\right)^{2/3} \times \frac{(2-v) \times (1+v)}{(2 \times (1-v^2))^{1/3}} \tag{7.8}$$

where R is the sphere radius, E is the elastic modulus, and v is the Poisson coefficient of the material.

It is interesting to note that PS to GS transition is proportional to the coefficient of friction and nonlinearly proportional to the normal load (to the power 2/3). Note that combining Eq. (7.1), the corresponding measured PS to GS transition is given as:

$$\delta_{m,t}^* = \delta_t^* + C_A \times Q_t^* = \delta_t^* + C_A \times \mu \times P \tag{7.9}$$

This effectively leads to a linear evolution if the test compliance is high.

Then, from the RCFM, the material response fretting maps (MRFM) can be established [7, 15–17]. Four domains are observed (Fig. 7.7). For very small displacement amplitudes inducing very small tangential force amplitude no damage occurs (I). Above a threshold displacement related to a threshold cyclic tangential force, crack nucleation can be observed (II). When the mixed fretting regime is achieved, cracking is still predominant but surface wear starts (III) inducing competition between the damage processes. Then, when gross-slip regime is established, fretting damage will be essentially governed by wear induced by debris formation and debris ejection (IV). The fretting map approach is a very useful strategy to compare materials as well as surface treatments and palliatives against fretting wear phenomena but remains a semiquantitative approach.

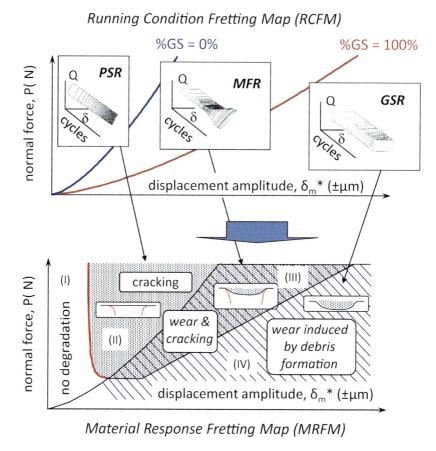

Fig. 7.7 Illustration of the fretting map approach.
Based on S. Fouvry, P. Kapsa, V. Vincent, Quantification of fretting damage. Wear 200 (1996) (1–2) 186–205.

7.4 Wear rate description

7.4.1 Archard and friction energy wear approach

Wear is a complex problem and many strategies were developed during the past decades to quantify this phenomenon in order to predict the wear volume extension. The usual strategy to quantify the wear rate in tribology consists in comparing the wear volume (V) extension as a function of the Archard parameter [18]. The Archard loading parameter is defined as the product of the normal force by the total sliding distance. For gross slip fretting sliding, it is expressed as:

$$\Sigma W = \sum_{i=1}^{N} 4 \times \delta_S(i) \times P(i) \qquad (7.10)$$

Fretting wear

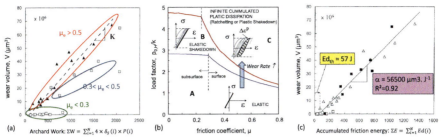

Fig. 7.8 (A) Analysis of the fretting wear extension of a sintered steel using Archard wear approach [19]; (B) K.L. Johnson displaying the various elastic and elastoplastic responses of metals as a function of the maximum contact pressure and the friction coefficient (sphere-on-flat contact) [9]; (C) Analysis of the fretting wear extension from (A) using the friction energy wear approach [19].

For constant loading conditions it is simplified to:

$$\Sigma W = 4 \times \delta_S \times P \times N \quad (7.11)$$

Then, plotting the evolution of the wear volume V versus ΣW (Fig. 7.8A), linear correlations are approximated to determine the so-called wear coefficient K (mm^3/N m) (or specific wear rate) so that:

$$V = K \times \Sigma W \quad (7.12)$$

One limitation of the Archard approach is that it does not consider the coefficient of friction in its formulation. Hence, under varying friction response a significant dispersion of the wear rate can be observed: the higher the coefficient of friction the higher the wear rate [19]. This can be interpreted using the K.L. Johnson diagram [9] where the elastic (A), elastic shakedown (B), and plastic shakedown or ratcheting metal responses (C) are reported as a function of the friction coefficient and the applied contact pressure (Fig. 7.8B). After this approach, as long as the metal stabilizes under elastic response (A and B), the cumulated plastic deformation dissipation is zero or negligible. Alternatively, if it doesn't stabilize (C), the plastic deformation is continuously increasing promoting high wear rates.

An alternative strategy to quantify wear consists in considering the accumulated friction energy (i.e., accumulated friction work) (Fig. 7.8C) [6, 7, 19].

$$\Sigma Ed = \sum_{i=1}^{N} Ed(i) \quad (7.13)$$

This approach is consistent with fretting analysis as it only requires summing the fretting loop area over the whole test duration. If the friction coefficient is constant, both Archard and friction energy approaches are equivalent so that:

$$\Sigma Ed = \mu_e \times \Sigma W = 4 \times \mu_e \times \delta_S \times P \times N \quad (7.14)$$

The wear volume extension is then quantified using an α energy wear coefficient:

$$V = \alpha \times \Sigma Ed \qquad (7.15)$$

Fig. 7.9 illustrates how for a given wear mechanism (abrasive wear of ceramic material), and a restricted spectrum of loading conditions, the energy approach can predict the wear volume. In Fig. 7.9A, various hard coatings are compared displaying linear evolutions and consequently reliable estimations of the energy wear coefficients [19–22]. Alternatively, Fig. 7.9B confirms the stability of the friction energy wear

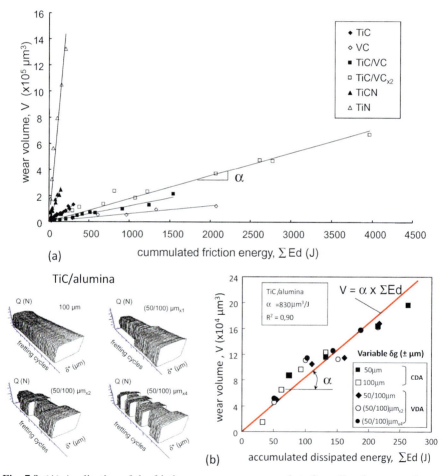

Fig. 7.9 (A) Application of the friction energy wear approach to formalize the wear volume extension of various Hard coating for constant gross slip condition [19]; (B) Illustrating how the friction energy wear approach is able to formalize the wear volume extension of a TiC/Alumina coating for constant (CDA) but also variable (VDA) gross slip displacement amplitudes (TiC/alumina) in a sphere-on-flat contact.
Based on S. Fouvry, T. Liskiewicz, C. Paulin, A global–local wear approach to quantify the contact endurance under reciprocating-fretting sliding conditions, Wear 263 (1–6) (2007) 518–531.

approach to predict the fretting wear volume extension of TiC/alumina hard coating interface even for complex varying gross slip-sliding conditions [19].

However, for the metallic interface, before generating wear debris, a plastic transformation is activated within the interface. This promotes a plastic recrystallization process inducing the formation of the so-called tribological transformed structure (TTS) which is very hard and very brittle [23–26].

This initial transformation obviously consumes a threshold friction energy Ed_{th} which can explain the energy offset usually observed in the metal wear volume—friction energy graph (Fig. 7.8C) [23].

The wear law needs then to be corrected by introducing an energy offset in the formulation [19,23] (Fig. 7.10A):

$$V = \alpha \times (\Sigma Ed - Ed_{th}) \tag{7.16}$$

where Ed_{th} is the threshold incubation energy related to the transformation of the metal interface into TTS (Fig. 7.10A).

As expected hard ceramic interface like TiC and TiN coating will display the lowest α energy wear coefficients (i.e., the highest wear resistances) as well as the lowest threshold incubation energies (Ed_{th}) since no plasticity can be activated (Fig. 7.10B).

7.4.2 Third-body theory

Archard and friction energy wear approaches are consistent as long as a similar wear process is activated. However, in many situations, the wear volume extension cannot be described assuming a single wear coefficient. Indeed, like any other tribological phenomena, fretting wear is a very complex process that not only depends on the contact materials but also the contact system. A key aspect concerns the debris layer also called the third body or debris bed. Entrapped within the interface, it highly influences both friction and wear processes (Fig. 7.11). This is particularly evident for fretting since small sliding amplitudes tend to maintain the wear debris within the interface. Many investigations have been performed during the past decades to better quantify this aspect introducing the third-body theory (TBT) [27–29]. According to this theory, the wear rate evolution is not controlled by the friction work inputted in the interface but related to the balance between the debris formation flow ϕ_f generated from the degradation of the first body (bulk materials) and the debris ejection flow ϕ_e. This approach implies an initial transition period during which the debris layer is formed and then, when this latter is stabilized, the wear rate stabilizes to a steady-state response. The steady-state wear rate regime is established when wear volume flow ϕ_V is equal to the debris ejection flow which itself is also equal to the debris formation flow [28].

$$\phi_V = \phi_f = \phi_e \tag{7.17}$$

Fig. 7.11 illustrates the principle of the third-body theory where both debris formation and debris ejection flows are plotted versus the debris layer thickness: the thicker the debris layer (h_{tb}) the larger the debris ejection flow during each sliding sequence.

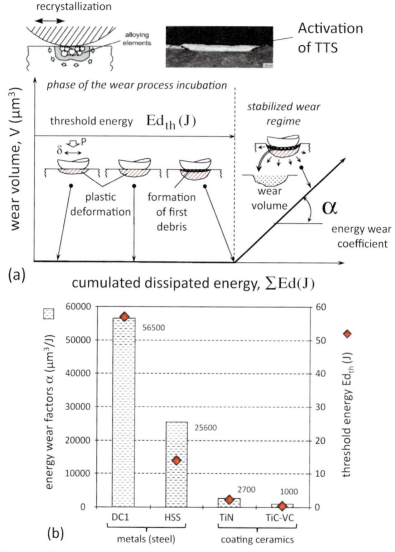

Fig. 7.10 (A) Illustration of the fretting wear process related to metal interfaces (incubation period related to the formation of TTS); (B) Comparison between metal and ceramic coating fretted against an alumina ball for similar loading conditions.
(A, B) Based on S. Fouvry, P. Kapsa, An energy description of hard coating wear mechanisms, Surf. Coat. Technol. 138 (2001) (2) 141–148.

Alternatively, the thicker the debris layer, the higher the friction energy consumed by the debris bed, the less damaged the first bodies, and finally the lower the debris formation flow. This theory suggests that the steady-state wear regime corresponds to a constant debris layer thickness ($h_{tb,ss}$). Obviously, by modifying the contact condition another equilibrium point will be established inducing another steady-state debris thickness. It will be detailed later the potential interest of this approach to better interpret the wear rate fluctuations as a function of the applied fretting loadings, contact size, and contact geometry but also depending on the rheological properties of the debris layer [29–31].

7.4.3 Contact oxygenation concept

As detailed previously, the stability of the wear rate also depends on the stabilized wear mechanism. Fretting wear usually promotes either abrasive or adhesive wear phenomena. Abrasive wear implies a plowing process of the softer material by the harder asperities favoring the formation of oxide powder debris. It could be two-body abrasive wear when the harder counterpart is the first body which directly interacts with the softer counterpart [32,33]. However, it could be a three-body process when the abrasive wear is controlled by the harder oxide particles constituting the third-body layer. Abrasive wear favors debris ejection producing a huge quantity of oxide particles. It generally leads to rather high wear rates although no seizure phenomena are observed. In contrast, adhesive wear phenomena inducing severe metal-metal transfers usually trigger rather low fretting wear rates since small sliding amplitudes limit the ejection of cohesive metal debris. However, catastrophic seizures inducing interface blocking are commonly observed. In many fretting interfaces, both abrasive and adhesive wear phenomena are detected simultaneously. The partition between the abrasive and adhesive process was recently formalized using the so-called contact oxygenation concept (COC) [34,35] (Fig. 7.12) or equivalently air-distilling [4] or oxygen exclusion process [36].

According to this concept, adhesive wear process is activated when the available dioxygen partial pressure in the interface falls below a threshold value $P_{O_2, th}$. The dioxygen partial pressure is maximum at the contact borders (open air) but decreases toward the inner part of the contact due to the consumption of dioxygen molecules reacting with the fresh metal to form oxide debris. The interface dioxygen partial pressure profile can therefore be established as a balance between the dioxygen diffusion rate from the

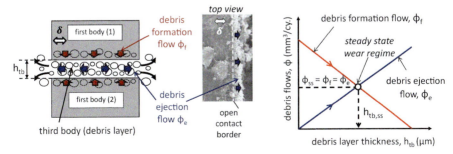

Fig. 7.11 Schematic presentation of the third body theory and the debris flow concept as a function of the debris layer thickness.
Based on N. Fillot, I. Iordanoff, Y. Berthier, Modelling, third body flows with a discrete element method—a tool for understanding wear with adhesive particles, Tribol. Int. 40 (2007) 973–981.

external borders and the reaction rate with the metal surface exposed by the friction work. This concept was recently modeled using an advection-dispersion-reaction model transposed by Baydoun et al. for fretting wear interfaces [37]. Solving the advection-dispersion-reaction equation by assuming porous properties of the debris layer and expressing the reaction parameter as a function of the "p.v" factor (i.e., pressure × sliding speed), it appears possible to estimate the dioxygen profile within the fretted interface and to predict the partition between the inner-adhesive and the abrasive external domains. On the other hand, using this approach allows predicting the extension of the lateral d_O distance from the contact borders (Fig. 7.12) below which debris oxidation and consequently abrasive wear are activated. Alternatively, the inner adhesive wear extension (d_A) is expressed as [35]:

$$d_A = d_C - d_O$$

where d_C is the distance from the inner to the external part of the contact along the axis where the analysis is established. The comparison between experiments and simulations for a flat-on-flat steel interface confirms the reliability of this approach. Note that

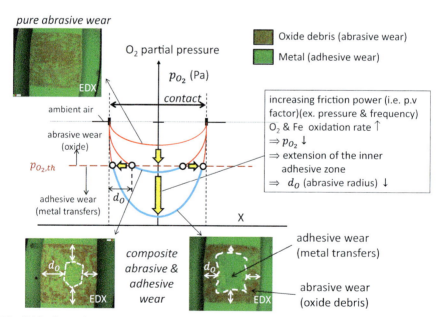

Fig. 7.12 Illustration of the contact oxygenation concept (COC) to formalize the transition from pure abrasive to composite abrasive-adhesive fretting scar interface (experimental validation using crossed flat-on-flat interface).
Based on S. Fouvry, P. Arnaud, A. Mignot, P. Neubauer, Contact size, frequency and cyclic normal force effects on Ti-6Al-4V fretting wear processes: an approach combining friction power and contact oxygenation. Tribol. Int. 113 (2017) 460–473; S. Baydoun, S. Fouvry, An experimental investigation of adhesive wear extension in fretting interface: application of the contact oxygenation concept, Tribo. Int. 147 (2020) 106266.

Fretting wear

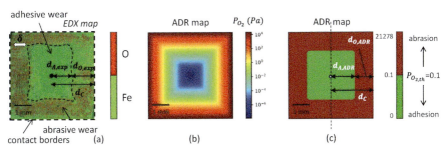

Fig. 7.13 (A) EDX mapping of a 35NCD16 flat-on-flat fretting contact (determination of the $d_{O,exp}$ and $d_{A,exp}$ length scales by comparing the concentration of oxygen versus iron); (B) Modeling O_2 partial pressure within the fretted interface using the ADR modeling (Advection-Dispersion-Reaction) [37]; (C) Prediction of the partition between abrasive and adhesive wear domains assuming $P_{O2,th} = 0.1$ Pa. A good correlation is observed with the EDX mapping observed in experiment (A).

adhesive wear in fretting was shown to be activated for an oxygen partial pressure not exceeding 0.1 Pa for a steel alloy tested under air-pressure controlled environments [38]. Based on this finding, it was assumed that $P_{O2,th} = 0.1$ Pa being the threshold value below which steel interface was observed to activate pure adhesive wear process (Fig. 7.13).

Like TBT, COC appears as a powerful approach to better formalize the transition from abrasive to adhesive wear processes in fretting interface. This aspect will be deepened later by investigating the influence of contact loadings.

7.5 Influence of contact loadings regarding wear rate fluctuations

As underlined previously, the energy wear rate is not constant and mainly depends on the contact loading conditions. Third body, contact oxygenation concept, as well as contact plasticity can significantly modify the wear process and consequently the fretting wear rate. To illustrate this aspect the effects of the siding amplitude, contact pressure, and sliding frequency will be examined.

7.5.1 Influence of the sliding amplitude

The sliding amplitude variable is already involved in the computation of the accumulated friction energy and therefore the energy wear rate would be expected to be constant. Nonetheless, numerous investigations highlighted an increase of the energy wear rate with the sliding amplitude particularly for adhesive wear materials [39]. This tendency can be explained considering the third-body theory assuming that the larger the sliding amplitude, the easier the debris ejection flow and consequently the better the energy wear rate efficiency (Fig. 7.14). To formalize this synergic effect,

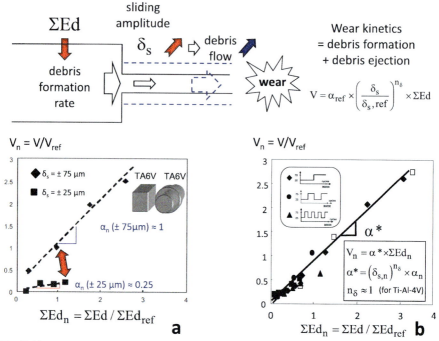

Fig. 7.14 Evolution of the fretting wear rate increase with the applied sliding amplitude for an adhesive wear Ti-6Al-4V interface [40]: (A) Basic friction energy approach; (B) Extended friction energy approach (results normalized versus a reference test condition; $\alpha_n = \alpha/\alpha_{Ref}$; $\delta_{s,n} = \delta_s/\delta_{s,Ref}$).

power-law function can be considered where the exponent expresses the sliding amplitude influence regarding the wear rate evolution (Fig. 7.14) [39].

7.5.2 Influence of the normal load

Like for the sliding amplitude, the wear rate coefficient would be assumed independent of the normal force since this variable is already involved in the formulation of the energy and Archard loading factors. The correlation of the wear extension versus the normal force rather than the contact pressure is justified by the Bowden and Tabor theory which suggests that the wear volume extension is indirectly proportional to the real contact area operating between asperities where plastic deformations of the contacting high spots (asperities) promote the formation of wear debris [41]. The real contact area (S) is directly proportional to the applied load (P) divided by the threshold contact pressure inducing plastic deformation (p_Y) which itself is approximately equal to $3\sigma_Y$, with Y being the yield stress in tension.

$$S = \frac{P}{p_y} = \frac{P}{3 \cdot \sigma_Y} \tag{7.18}$$

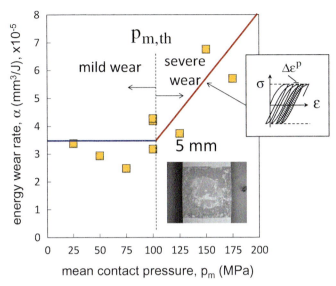

Fig. 7.15 Evolution of the wear rate of a flat-on-flat low-alloyed steel contact as a function of the mean contact pressure for gross-slip fretting sliding [42].

The investigation of a flat-on-flat low-alloyed steel interface (Fig. 7.15), allowing constant apparent contact area condition, confirmed that as long as the mean contact pressure remains lower than the threshold value (i.e., $p_{m,\,th} \approx 100$ MPa for the studied steel interface), the wear rate remains low and constant defining the so-called mild wear regime [42] as predicted by Archard and justified by Bowden and Tabor theories.

However, above a threshold pressure, a sharp increase of the wear rate is observed defining the so-called severe wear regime. As illustrated previously, the transition from mild to severe fretting wear rate can be interpreted using the K.L. Johnson diagram (Fig. 7.8B). Plastic shakedown or ratcheting is activated inducing severe plastic deformation and consequently a fast-rising of the wear rate.

7.5.3 Influence of the sliding frequency

Frequency is neither explicitly involved in the friction energy nor the Archard wear parameter. However, this parameter was extensively investigated during the past decades [41–43]. This research work underlines an increase in the wear rate when decreasing the sliding frequency. This rise was related to the formation and disruption of oxide films. The longer the sliding period, the thicker the oxide layer removed after each sliding sequence, and finally the higher the wear rate.

Various tribo-oxidation models suggest that the frequency dependency can be formalized using an Arrhenius formulation where the wear rate evolves as an inverse square root function of the frequency so that it could be expected an $f^{-0.5}$ wear rate dependency [44].

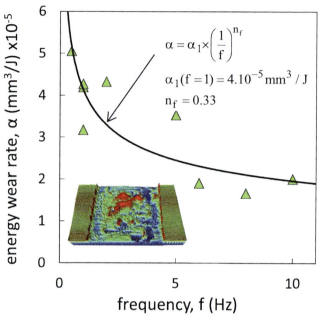

Fig. 7.16 Evolution of the energy wear rate as a function of the applied frequency (crossed flat 35NCD16 steel interface).
Based on S. Baydoun, S. Fouvry, S. Descartes, P. Arnaud, Fretting wear rate evolution of a flat-on-flat low alloyed steel contact: a weighted friction energy formulation, Wear 426 (2019) 676–693.

Some fluctuation of the frequency exponent has been however observed (i.e., $\alpha \propto f^{-0.33}$) (Fig. 7.16) which can be explained assuming the "contact oxygenation concept." Indeed, by increasing the sliding frequency and consequently the friction power dissipated in the interface, adhesive wear is extended and the global wear rate is consequently reduced because it does not only involve tribo-oxidation processes but also adhesive wear processes implying lower wear rates less dependent on the frequency.

7.5.4 Influence of the contact size

Recent investigations aim to evaluate how the contact size can influence the fretting wear rate. Keeping constant the contact pressure and the sliding amplitude, an asymptotic decreasing of the wear rate as a function of the contact size (i.e., half contact width) was detected (Fig. 7.17). This tendency was ascertained by showing that the fretting wear rate stabilizes above a threshold contact size. This was first explained by considering the third-body theory (TBT) such that by increasing the contact size, the delay to eject the debris is longer, the debris ejection flow is reduced and consequently, the wear rate is decreased. But more recently contact oxygenation concept (COC) provided an additional explanation, particularly to explain the asymptotic evolution. Indeed, by increasing the contact size, the inner adhesive area is extended and

Fretting wear

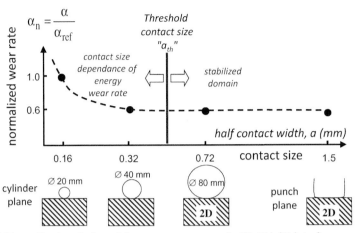

Fig. 7.17 Normalized evolution of the energy wear rate of a Ti-6Al-4V interface as a function of the contact size (keeping constant all the other fretting wear loading parameters).
Based on S. Fouvry, C. Paulin, S. Deyber, Impact of contact size and complex gross–partial slip conditions on Ti–6Al–4V/Ti–6Al–4V fretting wear, Tribol. Int. 42 (2009) (3) 461–474.

the global wear rate is reduced in consequence, since adhesive wear under fretting sliding tends to generate lower wear rates. When the adhesive/abrasive area partition stabilizes, the wear rate reaches a constant value [40]. One important conclusion is that small laboratory contact conditions tend to overestimate the fretting wear rate observed in large industrial assemblies providing a conservative prediction of the fretting wear damage.

7.5.5 Extended wear coefficient approach: A power law formulation

These different investigations underline that the energy wear rate of a given interface cannot be considered constant but depends on many factors. The evolution of the energy wear rate under fretting wear conditions can be interpreted qualitatively considering the third-body theory, the contact oxygenation concept, or the plastic response of the interface. A key question concerns the quantitative formulation of such fluctuation. It appears that a simple power-law function, can be derived applying a data optimization procedure involving a restricted number of experiments [42]:

$$V = \alpha^* \times \sum Ed$$

with

$$\alpha^* = \alpha_{ref} \cdot \left\{ \left(\frac{N}{N_{ref}}\right)^{n_N} \times \left(\frac{p}{p_{ref}}\right)^{n_p} \times \left(\frac{\delta_s}{\delta_{S_{ref}}}\right)^{n_{\delta_s}} \times \left(\frac{f}{f_{ref}}\right)^{n_f} \times \left(\frac{S}{S_{ref}}\right)^{n_S} \right\} \quad (7.19)$$

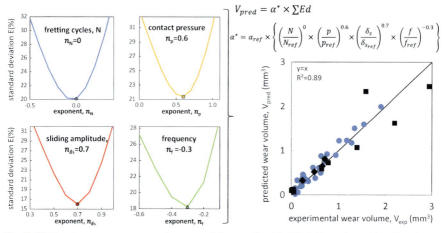

Fig. 7.18 Fretting wear response of a crossed flat-on-flat 35NCD16 interface: identification of the optimum exponents of the power function expressing the extended friction energy wear coefficient. Validation of the approach by comparing experimental and predicted wear volumes. Based on S. Baydoun, S. Fouvry, S. Descartes, P. Arnaud, Fretting wear rate evolution of a flat-on-flat low alloyed steel contact: a weighted friction energy formulation, Wear 426 (2019) 676–693.

where α^* is the extended energy wear coefficient, α_{ref} is the reference energy wear coefficient related to a key reference test condition, and n_N, n_p, $n_{\delta s}$, n_f, n_S are the exponent dependency related to fretting cycles (N), mean pressure (p), sliding amplitude (δ_s), frequency (f), and contact area (S).

Investigating the fretting wear response of a flat-on-flat 35NCD16 interface by varying the test duration, the contact pressure, the sliding amplitude and the frequency, Baydoun et al. obtained a rather accurate prediction of the total wear volume involving a very large spectrum of loading conditions (Fig. 7.18) [42].

Unfortunately due to the complexity of tribology processes, it is not possible at this stage, to establish an a priori formulation of the "extended" energy wear coefficient without a specific "calibration" from a selected number of experiments. Note that this analysis describes the general case of a dry metal/metal interface at ambient temperature. Numerous investigations underlined a huge fluctuation of the energy wear rate depending on the temperature, lubricated or water ambient conditions. Some aspects of these influences are briefly illustrated in the following sections.

7.6 Influence of the ambient conditions

7.6.1 Influence of temperature

Various fretting wear investigations underlined nonmonotonic evolution of the wear rate with the temperature particularly for Inconel or stainless steel (i.e., cobalt-based alloy) (Fig. 7.19). In the low and medium temperature range, the wear rate increases

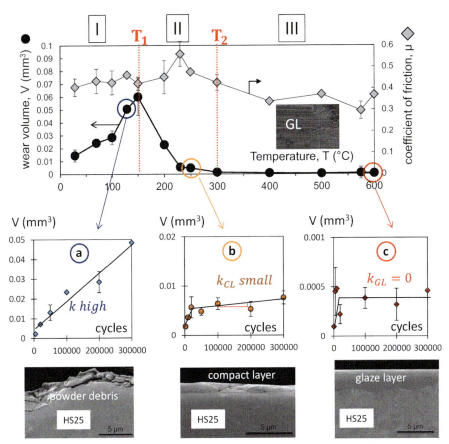

Fig. 7.19 Evolution of the fretting wear response of an HS25 Cobalt based alloy versus temperature. Activation of the lubricious and antiwear glaze layer above T2.
Based on A. Dreano, S. Fouvry, G. Guillonneau, Understanding and formalization of the fretting-wear behavior of a cobalt-based alloy at high temperature, Wear 452 (2020) 203297.

with the temperature due the synergic activation of the tribo-oxidation processes. However, above a threshold temperature, a tribosintering process of the debris layer is activated inducing the formation of a lubricious and above all protective third-body layer reducing the wear rate to nearly zero [45–47]. The process is reversible since when the temperature decreases below such a "glaze layer" threshold temperature (TGL), a linear high energy wear rate is observed again.

This peculiar behavior of the protective glaze layer was attributed to an overadhesive property of the oxide debris under high temperatures. The friction energy still induces the wear debris formation, but the wear debris particles are so adhesive that they are immediately embedded and sintered in the compact and agglomerated third-body layer. The debris cannot be ejected from the interface and finally, the global wear rate converges to zero. The "glaze layer" threshold temperature (TGL) was shown to highly depend on the metal composition. The higher is the concentration of the highly diffusing element, such as cobalt or iron, the lower the TGL and therefore the better the high-temperature wear resistance. A complete wear rate formulation as a function of the applied temperature is detailed in Ref. [47]. The authors demonstrated that using a restricted number of experiments and considering an Arrhenius—third-body hypothesis, it is possible to formalize such nonmonotonous evolution of the wear rate versus temperature.

7.6.2 Influence of the lubricated (grease) interface

Like temperature, lubricated interfaces under ambient conditions are mainly driven by the formation of specific tribofilms which in turn highly influence the wear rate evolution. Various authors [48–50] underlined a nonmonotonous evolution of the wear volume extension with the displacement amplitude which they suggested to be directly related to the friction behavior and lubricated tribofilm activation (Fig. 7.20). It was shown that as long as the fretting contact is running under partial slip, the lubricant is ejected from the interface and the coefficient of friction remains very high as observed for dry contact. Besides, because the friction energy under partial slip is very low, the wear volume is still negligible. This suggests that grease is not a pertinent palliative against the fretting cracking under partial slip condition. Note

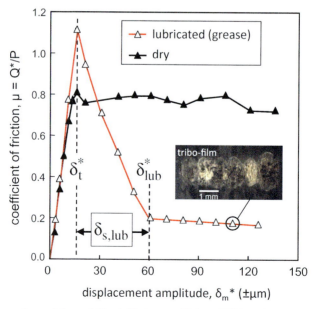

Fig. 7.20 Comparison of the stabilized friction coefficient (5000 cycles test duration) of a cylinder-on-flat smooth surface as a function of the applied measured displacement amplitude between dry and a grease-lubricated contact.
Based on S. Fouvry, V. Fridrici, C. Langlade, Ph. Kapsa, L. Vincent, Palliatives in fretting: a dynamical approach, Tribol. Int. 39 (2006) 1005–1015.

Fretting wear

that even higher tangential force amplitudes than in dry contacts can be observed under partial slip conditions due to the fact that the surrounding grease prevents the access of dioxygen molecules within the interface which reduces severe adhesive metal/metal interaction. However, above the partial slip/gross slip transition, a reduction of the coefficient of friction can be observed. This reduction of the coefficient of friction is not related to a "hydrodynamic" process as the fretting sliding speed is still very small (less than a few millimeter per second), but it can be explained by a "pumping" process which by injecting some lubricant within the fretted interface can promote the formation of lubricious tribofilms. The friction coefficient tends to decrease due to the sliding dissipation where a significant wear volume can be observed in this intermediate sliding domain above the gross slip transition. Increasing the displacement and therefore the sliding amplitude, promotes a faster formation of the lubricious protective tribofilm so that above a threshold lubricated sliding amplitude, the wear volume extension converges to nearly zero. The measured wear volume is here equal to the initial transient wear volume generated before the formation of the tribofilm. Surface analysis underlined that this protective tribofilm is very adhesive and consists of a mixture of lubricant and oxide wear debris. Various investigations underlined that the lower the viscosity of the oil, the faster the lubrication and consequently the better the fretting wear response of the greased lubricated interface.

7.6.3 Influence of the aqueous (corrosion) processes

The fretting wear response in water was extensively investigated by many researchers [51, 52] and more particularly by Mischler and coauthors [51–55] (Fig. 7.21). The latter developed a complete tribo-oxidation model allowing the prediction of the fretting wear volume as a function of the mechanical loading (friction energy) but also the electrical potential state imposed on the interface which indirectly monitors the

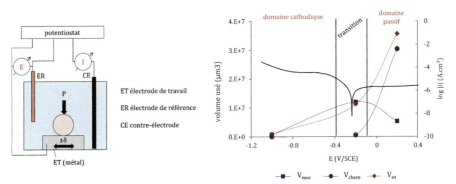

Fig. 7.21 Fretting wear rate under varying the potential state (highest wear rate under anodic domain).
Based on E. Marc, Analyse de la réponse tribologique d'un contact cylindre/plan soumis à des sollicitations de fretting sous chargement complexe: influence d'une solution Lithium-Bore (Thèse), University de Lyon, 2018, 2018LYSEC001.

corrosion wear volume. The model also depicts the synergic contribution involving coupled mechanical-corrosion processes. Two main aspects must be considered regarding the fretting wear response under aqueous conditions. First, the wear debris is systematically ejected from the interface (i.e., eliminated by the water flow) so that the third-body effect which is dominant under dry conditions can be neglected underwater conditions. However, as detailed by Mischler and co-authors, the presence of a corrosive environment by inducing a synergic effect between mechanical loadings and corrosion processes can dramatically increase the wear rate.

For instance, the development of offshore oil rigs provided many examples of fretting corrosion occurring on less corrosion-resistant materials. Hence, it was shown that the specific wear rate for a 1.5% Mn steel in the air is 1.07×10^{-16} m^3/N m, but in seawater, it increases up to 10.0×10^{-16} m^3/N m [56]. Fortunately, the application of cathodic protection of -950 mV versus standard calomel electrode (SCE) reduced the wear rate almost to that in air. Fig. 7.21 illustrates how the potential state imposed in fretting contact can modify the global wear rate underlining the necessity to apply cathodic protection to avoid dramatic fretting wear rate extension.

7.7 Surface wear modeling: Prediction of the maximum wear depth

7.7.1 Modeling of worn profiles

Modeling the fretting wear profiles was extensively investigated during the past decades [57–61]. Different strategies like semianalytical or boundary-element approaches could be adopted. However, for simple 2D contacts like the studied cylinder/plane interface, the Finite Element Method (FEM) appears as a convenient approach as it is simple to implement surface wear simulations in common commercial codes. The first developments regarding fretting wear were undertaken by McColl and coauthor by considering a local Archard wear approach [57]. It was successively transposed to the energy wear approach considering a bilateral wear surface analysis of the contacted interfaces [58]. The surface wear modeling (Fig. 7.22) consists in transposing the global wear volume approach at a local point of view where the total wear depth at a given x position is assumed proportional to the related accumulated friction energy density $\Sigma\varphi(x)$ through the corresponding energy wear coefficient (Fig. 7.22A):

$$h(x) = \alpha \times \Sigma\varphi(x) \tag{7.20}$$

Transposed to the studied cylinder/plane interface, the numerical procedure consists in computing, after a $\beta_{A,n}$ numerical fretting cycle, the increment of wear generated on plane and cylinder surfaces at the iteration n, respectively:

$$\Delta h_{p,n}(x) = \alpha_p \times \beta_{A,n} \times \frac{\varphi_{p/c,n}(x)}{2}$$

Fretting wear

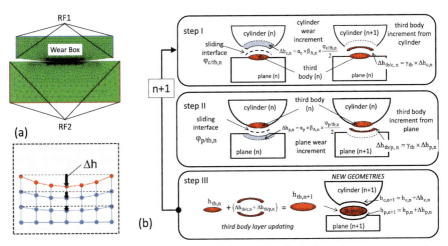

Fig. 7.22 Illustration of the surface wear modeling: (A) Cylinder/plan FEM interface and node translation procedure to update the worn profile after the *n*th iteration; (B) Description of the three-step procedure to update the plane and cylinder surfaces in addition to the third body layer.
Based on P. Arnaud, S. Fouvry, A dynamical FEA fretting wear modeling taking into account the evolution of debris layer, Wear 412 (2018) 92–108.

and

$$\Delta h_{c,n}(x) = \alpha_c \times \beta_{A,n} \times \frac{\varphi_{c/p,n}(x)}{2} \tag{7.21}$$

where $\varphi_{p/c,\,n}(x)/2$ is the friction energy density reported on the plane counterpart due to the friction between the plane and the cylinder and $\varphi_{c/p,\,n}(x)/2$ is the friction energy density reported on the cylinder due to the friction between the cylinder and the plane. The factor ½ is related to the fact that 50% of the total friction energy ($\varphi(x)$) is assumed reported to each counterface.

Plane and cylinder friction energy density distributions along the physical position *x* are in fact slightly different due to the relative sliding between the counterfaces (i.e., plane is assumed fixed and the cylinder moving). Obviously the integration of these two density profiles over the whole fretted interface leads to the same global friction energy (i.e., *Ed*).

These computations are long and fastidious, therefore, to avoid the simulation of each experimental fretting cycle, an acceleration $\beta_{A,\,n}$ factor is considered. This acceleration factor expresses the number of real (experimental) fretting cycles approximated by each numerical fretting cycle. The $\beta_{A,\,n}$ variable must be chosen large enough to allow fast simulation but small enough to avoid numerical distortions. The acceleration factor can be constant or adjusted as a function of the surface wear extension [58]. The contact geometry at the *N* numerical cycles is then updated by subtracting from the initial surface profile the computed worn thickness (h), so that a bilateral surface wear simulation could be achieved:

$$h_{p,n}(x) = h_{p,n-1}(x) - \Delta h_{p,n}(x)$$

$$h_{c,n}(x) = h_{c,n-1}(x) - \Delta h_{c,n}(x) \qquad (7.22)$$

Both plane and cylinder FEM geometries are then updated by applying a remeshing procedure. The final worn profiles are obtained when the number of simulated fretting cycles corresponds to target (i.e., experimental) fretting cycle (N) so that:

$$N = \sum_{n=1}^{N_{num}} \beta_{A,n} \qquad (7.23)$$

This approach provides rather good surface wear simulations when the fretting contact is not influenced by the presence of a debris layer as observed in lubrication (like water). However, in dry contact when the debris particles display high-cohesive properties, like for instance in Ti-6Al-4V interface, a thick third-body layer is formed modifying, in consequence, the contact pressure and accordingly the wear profile evolution. Many researchers intend to consider the presence of a debris layer in fretting wear simulation [60, 61]. However, these numerical developments are usually "static" in the sense that they do not evolve with the fretting cycles and surface damage. More recently, Arnaud et al. [62] proposed a dynamical description where the third-body layer is simulated as an additional FEM part entrapped between the plane and the cylinder counterparts (Fig. 7.23).

An originality of this approach lies in the fact that this third-body layer extends laterally and in thickness as a function of the surface wear extension. Hence, after the Nth numerical fretting cycle, a $\gamma_{tb,n}(x)$ proportion of the worn thickness increment is transposed to the third-body layer. On the other hand, the $h_{tb,n}(x)$ thickness of the debris layer at the x position and the nth computation iteration evolves so that:

$$h_{tb,n}(x) = h_{tb,n-1}(x) + \gamma_{tb,n}(x) \times (\Delta h_{p,n} + \Delta h_{c,n}) \qquad (7.24)$$

Note that the complement of the third-body transfer, [i.e., $1 - \gamma_{tb,n}(x)$] corresponds to the proportion of the worn debris eliminated from the interface. Regarding surface wear simulations, similar formulations are considered except in Eq. (7.21) where the presence of the third-body must be considered which implies involving two sliding interfaces between the third-body layer versus the top and the bottom counterparts, respectively. As discussed by Arnaud et al. [62], the third-body conversion factor "$\gamma_{tb,n}(x)$" expressing locally, at the nth fretting cycle, the proportion of the worn thickness transfer to the third-body layer appears as a key aspect of the surface wear modeling (Fig. 7.24). Regarding the surface distribution, an elliptic function leading to a maximum value at the center of the contact allows describing the fact that the worn debris generated at the center of the contact displays a higher probability to remain in the third body than the particles formed on the lateral sides where they are more easily ejected. Alternatively, the fretting cycle evolution of $\gamma_{tb,n}$ amplitude

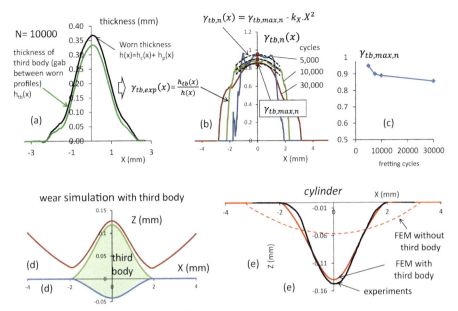

Fig. 7.23 Illustration of a simulated U-shape fretting scar taking into account the presence of the third-body layer for a dry Ti-6Al-4V cylinder/plane interface: (A) Comparison between the worn and the third-body thicknesses at 10,000 fretting cycles and the computation of the third-body conversion profile; (B) Formulation of the $Y_{tb,n}(\chi)$ third-body conversion parameter as a function of the lateral position and fretting cycles (i.e., surface wear extension); (C) Asymptotic decreasing of the $Y_{tb,max,n}$ coefficient with the wear extension; (D) Modeling the plane and the cylinder worn surfaces and the related third-body layer; (E) Comparison between the experiments and the simulations with and without third body. (Third body simulation is required to achieve a reliable prediction.)
Based on P. Arnaud, S. Fouvry, A dynamical FEA fretting wear modeling taking into account the evolution of debris layer, Wear 412 (2018) 92–108.

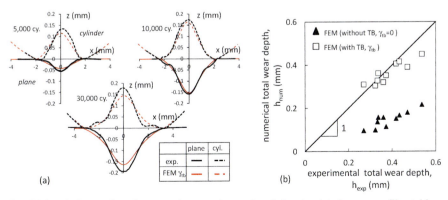

Fig. 7.24 (A) Comparison between the experimental and the simulated worn profiles taking into account the presence of the third body through the third-body conversion factor $Y_{tb,n}$; (B) Comparison between the experimental and the simulated maximum wear depth taking into account the presence of the debris layer.
Based on P. Arnaud, S. Fouvry, A dynamical FEA fretting wear modeling taking into account the evolution of debris layer, Wear 412 (2018) 92–108.

can be formalized using an asymptotic decreasing function from a maximum value around 1 at the beginning of the test when all the worn debris remains entrapped in the contact until a minimum value around zero when the steady-state third-body thickness is established. Hence, by considering the local friction energy wear approach and simulating the presence and the evolution of the third-body layer, it appears possible to predict the fretting wear profile evolutions as well as the maximum wear depth evolutions which are key aspects to quantify the fretting wear durability of industrial assemblies.

7.7.2 Prediction of the coating durability using the friction energy density parameter

Many studies were conducted during the past decades to improve the wear resistance of fretting interfaces applying soft, hard and solid lubricant coatings. These surface treatments reduce the coefficient of friction and limit seizure phenomenon. However, to predict the coating durability, an alternative strategy to wear volume analysis needs to be adopted (Fig. 7.25). Indeed, the coating durability is usually related to the critical number of loading cycles (Nc) when the sliding interface reaches the substrate interface. For solid lubricant coatings, Nc can be easily detected by measuring the discontinuous increase of the coefficient of friction. When the coating doesn't lead to significant fluctuation of the coefficient of friction, Nc can be estimated by performing interrupted tests and measuring the maximum wear depth. When this latter is longer than the coating thickness, this infers that the coating failure is achieved. The durability of a coating Nc can be expressed as a function of the applied displacement amplitude for a given normal load as detailed by Langlade et al. [63]. It can be also formalized as a function of the maximum friction energy density (or Archard density parameter) as detailed previously in Refs. [64,65]. Indeed, considering thin coating layers less than few micron thicknesses, the worn surface extension can be neglected and the initial maximum friction energy density appears as a reliable parameter to establish the fretting wear durability of coatings. Using this approach, the contact pressure and the sliding amplitude can be combined through a single-energy loading variable [66]. A power law formulation can be defined so that the coating endurance can be related to a given friction energy capacity (χ) characterizing the fretting performance of a given surface treatment (i.e., coating).

$$\text{Nc} = \frac{\chi}{\varphi_{max}} \tag{7.25}$$

where φ_{max} is the maximum friction energy density inputted in the contact during a fretting cycle given as:

$$\varphi_{max} = 4 \times \delta_s \times \mu_e \times p_{max} \tag{7.26}$$

where p_{max} is the maximum pressure value in the contact.

Fretting wear

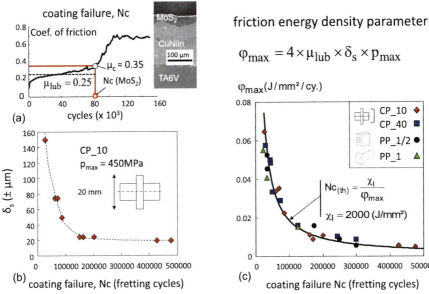

Fig. 7.25 Quantification of the coating durability under fretting wear (analysis of a MoS2 solid lubricant): (A) Coating failure Nc when the substrate is reached (friction discontinuity); (B) Evolution of the coating failure Nc as a function of δ_s; (C) Quantification of Nc as a function of the maximum local Archard work density for various contact geometries (*CP*, cylinder/plane; *PP*, punch/plane), sliding amplitude (δ_s) and maximum contact pressure (P_{max}). Based on S. Fouvry, C. Paulin, An effective friction energy density approach to predict solid lubricant friction endurance: application to fretting wear, Wear 319 (2014) 211–226.

An equivalent Archard's work description can be considered substituting in the equation the variable χ by the equivalent Archard's work capacity $\psi = \chi/\mu$ and φ_{max} by the corresponding Archard work density:

$$\omega_{max} = 4 \times \delta_s \times p_{max} \tag{7.27}$$

so that

$$\text{Nc} = \frac{\psi}{\omega_{max}} \tag{7.28}$$

For hard coating layers (Fig. 7.26), the friction energy capacity can be related to the energy wear coefficient (α) assuming a linear extension of the maximum wear depth versus the maximum accumulated friction energy density so that:

$$\chi = \frac{t_n}{\alpha} \tag{7.29}$$

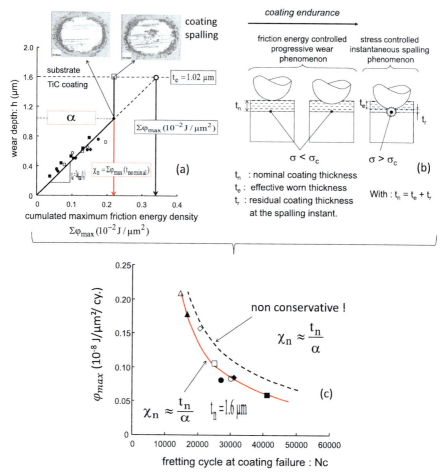

Fig. 7.26 Application of the friction energy density approach to formalize the fretting wear durability of thin TiC hard coating [64, 65]: (A) Evolution of the maximum wear depth as a function of the cumulated friction energy density. (B) Coating failure is occurring at $h = t_e$ due to brutal decohesion phenomenon so that an effective coating thickness (t_e) smaller than the nominal one (t_n) must be considered to rationalize the coating durability; (C) Formalization of the coating durability Nc as a function of the maximum friction energy density inputted in the interface φ_{max} through the effective friction energy capacity of the studied TiC coating (χ_e).

However, in many situations, the brutal spalling phenomenon inducing a coating failure before the fretted interface reaches the substrate reduces significantly the expected fretting endurance. This aspect can be formalized by considering the effective coating thickness (t_e) related to the worn thickness before the coating layer is definitively removed from the interface due to a cracking cohesion process so that:

$$\text{Nc} = \frac{\chi_e}{\varphi_{max}} \tag{7.30}$$

with

$$\chi_e = \frac{t_e}{\alpha} = \frac{t_n - t_r}{\alpha} \qquad (7.31)$$

where t_r is the residual coating thickness just before failure. Hence, using this basic approach it is possible to formalize the fretting wear endurance of hard coatings like TiC.

7.8 Conclusion

This chapter illustrates some aspects regarding the quantification of the fretting wear damage. First, the sliding condition characterized by the tangential force-displacement ($Q^* - \delta_m$) fretting loop is quantified using various sliding ratios. Using such scalar variables, it is possible to establish if the contact is running under partial or gross slip condition. Then, using the GS% ratio expressing the proportion of a gross-slip sliding during a fretting test, it is possible to formalize the sliding regimes. Introducing the fretting map approach, it was shown that the partial slip regime mainly favors cracking nucleation and propagation whereas larger displacement conditions inducing gross slip regime mainly promote surface wear by debris formation and ejection. Intermediate displacement conditions inducing mixed fretting regimes display competition between wear and cracking phenomena. This chapter only focuses on surface wear mechanisms therefore only gross slip regime condition was investigated.

The quantification of the fretting wear extension under gross slip condition could be addressed using the Archard work approach (i.e., product of the normal force by the total sliding distance) but contact mechanics underlines a huge effect of the friction coefficient regarding contact stresses. Hence, the friction energy wear approach, expressing the wear volume as a linear function of the accumulated friction work multiplied by the energy wear coefficient (α), appears more reliable. However, many investigations underline that the energy wear coefficient (α) is not constant but depends on the wear mechanisms involved in the fretted interface which itself depends on the loading conditions. To interpret such fluctuations both third-body theory (TBT) and contact oxygenation concept (COC) must be considered. Third-body theory (TBT) suggests that the wear kinetics is a function of the balance between the debris formation and debris ejection flows. If the debris ejection flow is reduced, for instance, because the third-body layer is very cohesive and the debris cannot be ejected, then the global wear rate is sharply reduced. This phenomenon was observed for high-temperature test conditions when a very cohesive glaze layer is formed. Alternatively, many fretting wear investigations underlined the formation of a composite adhesive-abrasive fretting wear interface. Because different wear coefficients are operating if the adhesive or abrasive wear is occurring, a significant fluctuation of the global energy wear coefficient could be observed depending on the relative distribution of abrasive or adhesive domains. This composite interface distribution can be formalized using the so-called contact oxygenation concept (COC) which consists in computing the dioxygen partial pressure profile within the fretted interface such that the extension of the inner adhesive wear domain can be expected whenever $P_{O2}(x,y) < P_{O2, th}$.

Using both TBT and COC approach, it was then possible to interpret the effect of the contact loading conditions regarding the wear coefficient fluctuation. Hence, it was shown that an increase of the sliding amplitude, by favoring the debris ejection flow, tends to increase the wear coefficient. Besides, increasing the normal force, by inducing severe plastic deformations, also increases the wear rate coefficient. Alternatively, an increase of the sliding frequency, by reducing the contact oxygenation, favors adhesive wear which by displaying a lower wear coefficient than abrasive one tends to decrease the global wear rate coefficient. Finally, both TBT and COC approaches can explain the general trends regarding the contact size effect such that larger contact sizes induce smaller wear rates until an asymptotic value is reached. Indeed, increasing the contact size will simultaneously reduce the debris ejection flow due to the longer debris travel before ejection in addition to the contact oxygenation due to the longer distance of the inner part from the open air. An a priori formulation expressing the influence of each of these loading parameters is still under development. However, recent developments suggest interesting multiphysics approaches. An alternative strategy, based on an adequate data processing, consists in expressing an extended energy wear coefficient through a power law formulation as a function of the sliding amplitude, contact pressure, frequency and contact area. Despite its simplicity, rather nice correlations with experiments were observed. The prediction of the fretting wear volume is an important aspect but in many industrial applications a key aspect concerns the modeling of the maximum wear depth extension. This implies the transposition of the global "wear volume-cumulated friction energy" approach to a local "wear depth-cumulated friction energy density" description. Different numerical strategies can be considered like for instance finite element simulations. Various investigations suggested that achieving a reliable prediction of the maximum wear depth prediction necessitates considering the presence of the debris layer and its evolution with the surface wear extension. Finally, it is shown that using a local friction energy density approach, the maximum wear depth as well as the surface coating durability can be predicted. This approach was successfully adopted to estimate the fretting wear durability of solid lubricant by expressing the Nc fretting wear endurance as an inverse function of the maximum friction energy density (φ_{max}) inputted in the interface assuming a "friction energy capacity" (χ) characterizing the performance of the studied surface treatment (i.e., Nc $=\chi/\varphi_{max}$). A similar correlation could be derived for hard coatings such that the "friction energy capacity" can be expressed as function of the energy wear rate and the coating thickness. However, in many situations, brutal coating failure occurs before the fretted interface reaches the substrate so that an effective coating thickness (i.e., $t_e = h$ at the coating failure) must be considered to estimate the coating performance such that $\chi \approx t_e/\alpha$. This brief description illustrates how fretting wear damage can be formalized using a mechanical friction energy description. As detailed in this book, corrosion and tribocorrosion phenomena can significantly modify such fretting wear processes underlining again the complexity of fretting wear degradations like any tribological surface damage process.

7.9 Notations

Latin letters

A	friction energy sliding criterion
d_O	distance from the interface boundaries to the frontiers of the adhesion zone (abrasive wear extension) [mm]
d_A	width of the inner part of the contact enduring adhesive wear [mm]
f	frequency [Hz]
$GS\%$	fretting regime index (i.e., relative % of gross slip sliding during a test)
h	total wear depth (i.e., sum of the two counterfaces) [µm]
$h(x)$	total wear depth profile (i.e., sum of the two counterfaces) at the x position [µm]
$h_{c,\,n}(x)$	cylinder worn profile at the nth fretting cycle
h_{max}	total maximum wear depth (i.e., sum of the two counterfaces) [µm]
$h_{p,\,n}(x)$	plane worn profile at the nth fretting cycle [µm]
$h_{tb,\,n}(x)$	thickness of the third-body layer at the nth fretting cycle [µm]
K	Archard wear coefficient [mm^3/(N m)]
N	applied fretting cycles
N_C	fretting cycle related to the coating failure
p	contact pressure [MPa]
P	normal force [N]
p_m	mean contact pressure [MPa]
p_{max}	maximum contact pressure [MPa]
P_{O2}	dioxygen partial pressure within the interface [Pa]
Q	tangential force [N]
Q^*	tangential force amplitude [N]
S	apparent contact area [mm^2]
T	temperature [°C]
t	time [s]
t_n	nominal coating thickness [µm]
t_e	effective coating thickness [µm]
V	wear volume [mm^3 or µm^3]

Greek letters

α	total (global contact) energy wear coefficient [mm^3/J]
α_c	energy wear coefficient of cylinder surface [mm^3/J]
α_p	energy wear coefficient of plane surface [mm^3/J]
$\beta_{A,\,n}$	number of real fretting cycles simulated during the nth numerical fretting cycle (acceleration factor)
$\gamma_{tb,\,n}$	third-body conversion factor (proportion of the worn thickness transferred to the third-body layer at the nth fretting cycle)
δ	effective contact displacement [µm]

δ_A	displacement accommodated by the test apparatus [µm]
δ_m	measured displacement [µm]
δ_m^*	measured displacement amplitude [µm]
$\delta_{m,t}^*$	measured displacement amplitude at the gross slip transition [µm]
δ_s	effective sliding amplitude [µm]
$\delta_{s,lub}$	effective sliding amplitude inducing lubrication in grease interface [µm]
δ_t^*	effective contact displacement at the sliding transition [µm]
$\Delta h_{p,n}$	increment of the wear depth generated on the plane surface during the nth cycle [µm]
$\Delta h_{c,n}$	increment of the wear depth generated on the cylinder surface during the nth cycle [µm]
μ	nominal friction coefficient
μ_e	energy friction coefficient
ΣW	accumulated Archard work [N m]
$\Sigma \varphi(x)$	accumulated friction energy density dissipated at the x position [J/mm^2]
ΣEd	accumulated friction energy (i.e., friction work) [J]
σ_Y	plastic yield stress [MPa]
ϕ	debris flow
$\varphi(x)$	friction energy density dissipated at the x position during a fretting cycle [J/mm^2]
φ_{max}	maximum friction energy density dissipated in the interface during a fretting cycle [J/mm^2]
χ	friction energy capacity [mm^2/J]
ψ	Archard's work capacity [mm^2/(N m)]
ω	Archard work density per fretting cycle [N m/mm^2]
ω_{max}	maximum Archard work density in the interface per fretting cycle [N m/mm^2]

References

[1] D. Hoeppner, Mechanisms of Fretting Fatigue and Their Impact On Test Methods Development, ASTM STP 1159, 1992, pp. 23–32.
[2] R.B. Waterhouse, Fretting wear, Wear 100 (1–3) (1984) 107–118.
[3] M. Kuno, R.B. Waterhouse, D. Nowell, D.A. Hills, Initiation and growth of fretting fatigue cracks in the partial slip regime, Fatigue Fract. Eng. Mater. Struct. 12 (1989) 387–398.
[4] C. Mary, S. Fouvry, J.M. Martin, B. Bonnet, Pressure and temperature effects on fretting Wear damage of a Cu–Ni–In plasma coating versus Ti17 titanium alloy contact, Wear 272 (1, 3) (2011) 18–37.
[5] Z.R. Zhou, S. Fayeulle, L. Vincent, Cracking behaviour of various aluminium alloys during fretting wear, Wear 155 (1992) 317–330.
[6] H. Mohrbacker, B. Blanpain, J.P. Celis, J.R. Roos, L. Stals, M. Van Stappen, Oxidational wear of TiN coatings on tool steel and nitride tool steel in unlubricated fretting, Wear 188 (1995) 130–137.
[7] S. Fouvry, P. Kapsa, V. Vincent, Quantification of fretting damage, Wear 200 (1–2) (1996) 186–205.
[8] R.D. Mindlin, H. Deresiewicz, Elastic spheres in contact under varying oblique forces, Trans. ASME Ser. E J. Appl. Mech. 20 (1953) 327–344.
[9] K.L. Johnson, Contact Mechanics, Cambridge University Press, Cambridge, 1985.
[10] S. Fouvry, P. Kapsa, L. Vincent, Analysis of sliding behaviour for fretting loadings: determination of transition criteria, Wear 185 (1995) 35–46.

[11] M. Varenberg, I. Etsion, G. Halperin, Slip index: a new unified approach to fretting, Tribol. Lett. 17 (3) (2004) 569–573.
[12] O. Vingsbo, S. Söderberg, On fretting maps, Wear 126 (1988) 131–147.
[13] Z. Zhou, Mixed fretting regime, Wear 181 (1995) 531–536.
[14] S. Heredia, S. Fouvry, Introduction of a new sliding regime criterion to quantify partial, mixed and gross slip fretting regimes: correlation with wear and cracking processes, Wear 269 (7–8) (2010) 515–524.
[15] O. Vingsbo, M. Odfalk, N.-E. Shen, Fretting maps and fretting behaviour of some FCC metal alloys, Wear 13 (1990) 153–167.
[16] Z.R. Zhou, K. Nakazawa, M.H. Zhu, N. Maruyama, P. Kapsa, L. Vincent, Progress in fretting maps, Tribol. Int. 39 (2006) 1068–1073.
[17] C. Colombier, Y. Berthier, A. Floquet, L. Vincent, M. Godet, Fretting: load carrying capacity of wear debris, J. Tribol. 106 (1984) 194–201.
[18] J.F. Archard, Contact and rubbing of flat surfaces, J. Appl. Phys. 24 (8) (1953) 981.
[19] S. Fouvry, P. Kapsa, An energy description of hard coating wear mechanisms, Surf. Coat. Technol. 138 (2) (2001) 141–148.
[20] S. Achanta, D. Drees, J.P. Celis, Friction and nanowear of hard coatings in reciprocating sliding at milli-Newton loads, Wear 259 (2005) 719–729.
[21] Y. Liu, T. Liskiewicz, A. Yerokhin, A. Korenyi-Both, J. Zabinski, M. Lin, A. Matthews, A. Voevodin, Fretting wear behavior of duplex PEO/chameleon coating on Al alloy, Surf. Coat. Technol. 352 (2018) 238–246.
[22] S. Fouvry, T. Liskiewicz, C. Paulin, A global–local wear approach to quantify the contact endurance under reciprocating-fretting sliding conditions, Wear 263 (1–6) (2007) 518–531.
[23] E. Sauger, S. Fouvry, L. Ponsonnet, P. Kapsa, J.M. Martin, et al., Tribologically transformed structure in fretting, Wear 245 (1) (2000) 39–52.
[24] V. Nurmi, J. Hintikka, J. Juoksukangas, M. Honkanen, M. Vippola, A. Lehtovaara, et al., The formation and characterization of fretting-induced degradation layers using quenched and tempered steel, Tribol. Int. 131 (2019) 258–267.
[25] A.M. Kirk, P.H. Shipway, W. Sun, C.J. Bennett, The effect of frequency on both the debris and the development of the tribologically transformed structure during fretting wear of a high strength steel, Wear 426 (2019) 694–703.
[26] D. Tumbajoy-Spinel, S. Descartes, J.-M.L. Bergheau, V. Lacaille, G. Guillonneau, J. Michler, G. Kermouche, Assessment of mechanical property gradients after impact-based surface treatment: application to pure α-iron, Mater. Sci. Eng. A667 (2016) 189–198.
[27] M. Godet, Third-bodies in tribology, Wear 136 (1990) 29–45.
[28] N. Fillot, I. Iordanoff, Y. Berthier, Modelling, third body flows with a discrete element method—a tool for understanding wear with adhesive particles, Tribol. Int. 40 (2007) 973–981.
[29] S. Descartes, Y. Berthier, Rheology and flows of solid third bodies: background and application to an $MoS_{1.6}$ coating, Wear 252 (2002) 546–556.
[30] J.D. Lemm, A.R. Warmuth, S.R. Pearson, P.H. Shipway, The influence of surface hardness on the fretting wear of steel pairs-its role in debris retention in the contact, Tribol. Int. 81 (2015) 258–266.
[31] A. Viat, G. Guillonneau, S. Fouvry, G. Kermouche, S. Sao Joao, J. Wehrs, J. Michler, J.F. Henne, Brittle to ductile transition of tribomaterial in relation to wear response at high temperatures, Wear 392–393 (2017) 60–68.
[32] K. Holmberg, A. Matthews, Properties, Mechanisms, Techniques and Applications in Surface Engineering, Elsevier Science, 2009, p. 576, ISBN: 9780444527509.

[33] G.W. Stachowiak, Wear, Materials, Mechanics and Practice, Tribology in Practice Series, Wiley, 2006, p. 458.
[34] S. Fouvry, P. Arnaud, A. Mignot, P. Neubauer, Contact size, frequency and cyclic normal force effects on ti-6al-4v fretting wear processes: an approach combining friction power and contact oxygenation, Tribol. Int. 113 (2017) 460–473.
[35] S. Baydoun, S. Fouvry, An experimental investigation of adhesive wear extension in fretting interface: application of the contact oxygenation concept, Tribol. Int. 147 (2020) 106266.
[36] A.R. Warmuth, S.R. Pearson, P.H. Shipway, W. Sun, The effect of contact geometry on fretting wear rates and mechanisms for a high strength steel, Wear 301 (1–2) (2013) 491–500.
[37] S. Baydoun, P. Arnaud, S. Fouvry, Modelling adhesive wear extension in fretting interfaces: an advection-dispersion-reaction contact oxygenation approach, Tribol. Int. 151 (2020) 106490.
[38] A. Iwabuchi, T. Kayaba, K. Kato, Effect of atmospheric pressure on friction and wear of 0.45% C steel in fretting, Wear 91 (1983) 289–305.
[39] S. Fouvry, P. Duo, P. Perruchaut, A quantitative approach of Ti-6Al-4V fretting damage: friction, wear and crack nucleation, Wear 257 (9–10) (2004) 916–929.
[40] S. Fouvry, C. Paulin, S. Deyber, Impact of contact size and complex gross–partial slip conditions on Ti–6Al–4V/Ti–6Al–4V fretting wear, Tribol. Int. 42 (3) (2009) 461–474.
[41] F.P. Bowden, D. Tabor, The Friction and Lubrication of Solids, Oxford University Press, 1950.
[42] S. Baydoun, S. Fouvry, S. Descartes, P. Arnaud, Fretting wear rate evolution of a flat-on-flat low alloyed steel contact: a weighted friction energy formulation, Wear 426 (2019) 676–693.
[43] R.B. Waterhouse, in: D. Scott (Ed.), Fretting, Treatise on Materials Science and Technology, Academic Press, 1978, pp. 259–286.
[44] A. Dreano, S. Fouvry, G. Guillonneau, A tribo-oxidation abrasive wear model to quantify the wear rate of a cobalt based alloy subjected to fretting in low-to-medium temperature conditions, Tribol. Int. 125 (2018) 128–140.
[45] F.H. Stott, G.C. Wood, The influence of oxides on the friction and wear of alloys, Tribol. Int. 11 (1978) 211–218.
[46] X. Jin, P.H. Shipway, W. Sun, The role of temperature and frequency on fretting wear of a like-on-like stainless steel contact, Tribol. Lett. 65 (2017) 77.
[47] A. Dreano, S. Fouvry, G. Guillonneau, Understanding and formalization of the fretting-wear behavior of a cobalt-based alloy at high temperature, Wear 452 (2020) 203297.
[48] M. Shima, H. Suetake, I.R. McColl, R.B. Waterhouse, M. Takeuchi, On the behaviour of an oil lubricated fretting contact, Wear 210 (1997) 304–310.
[49] Z.R. Zhou, L. Vincent, Lubrification in fretting a review, Wear 225 (9) (1999) 962–968.
[50] S. Fouvry, V. Fridrici, C. Langlade, P. Kapsa, L. Vincent, Palliatives in fretting: a dynamical approach, Tribol. Int. 39 (2006) 1005–1015.
[51] J. Geringer, D.D. Macdonald, Friction/fretting-corrosion mechanisms: current trends and outlooks for implants, Mater. Lett. 134 (2014) 152–158.
[52] P. Ponthiaux, F. Wenger, D. Drees, J.P. Celis, Electrochemical techniques for studying tribocorrosion processes, Wear 256 (2004) 459–468.
[53] A. Dalmau, W. Rmili, C. Richard, A. Igual-Muñoz, Tribocorrosion behavior of new martensitic stainless steels in sodium chloride solution, Wear 368 (2016) 146–155.
[54] S. Mischler, Triboelectrochemical techniques and interpretation methods in tribocorrosion: a comparative evaluation, Tribol. Int. 41 (2008) 573–583.

[55] S. Mischler, A.I. Munoz, Tribocorrosion, in: K. Wandelt (Ed.), Encyclopedia of Interfacial Chemistry, Elsevier, New York, 2018, pp. 504–514.
[56] B.R. Pearson, R.B. Waterhouse, The fretting corrosion in seawater of materials used in offshore structures, in: Proc. 9th International Congress on Metallic Corrosion, Toronto, June, vol. 2, 1984, pp. 334–341.
[57] I.R. McColl, J. Ding, S.B. Leen, Finite element simulation and experimental validation of fretting wear, Wear 256 (2004) 1114–1127.
[58] C. Mary, S. Fouvry, Numerical prediction of fretting contact durability using energy wear approach: optimisation of finite-element model, Wear 263 (1–6) (2007) 444–450.
[59] L. Gallego, B. Fulleringer, S. Deyber, D. Nelias, Multiscale computation of fretting wear at the blade/disk interface, Tribol. Int. 43 (4) (2010) 708–718.
[60] S. Basseville, G. Cailletaud, An evolution of the competition between wear and crack initiation in fretting conditions for Ti-6Al-4V alloy, Wear 328 (2015) 443–455.
[61] T. Yue, A.L. Wahab, Finite element analysis of fretting wear under variable coefficient of friction and different contact regimes, Tribol. Int. 107 (2017) 274–282.
[62] P. Arnaud, S. Fouvry, A dynamical FEA fretting wear modeling taking into account the evolution of debris layer, Wear 412 (2018) 92–108.
[63] C. Langlade, B. Vannes, M. Taillandier, M. Pierantoni, Fretting behavior of low-friction coatings: contribution to industrial selection, Tribol. Int. 34 (2001) 49–56.
[64] S. Fouvry, C. Paulin, T. Liskiewicz, Application of an energy wear approach to quantify fretting contact durability: introduction of a wear energy capacity concept, Tribol. Int. 40 (10–12) (2007) 1428–1440.
[65] T. Liskiewicz, S. Fouvry, Development of a friction energy capacity approach to predict the surface coating endurance under complex oscillating sliding conditions, Tribol. Int. 38 (1) (2005) 69–79.
[66] S. Fouvry, C. Paulin, An effective friction energy density approach to predict solid lubricant friction endurance: Application to fretting wear, Wear 319 (2014) 211–226.

Harmonic analysis of tribocorrosion: Identification of repassivation kinetics and separation of reactive (corrosion) and mechanical (wear) contributions

8

Michel Keddam and Vincent Vivier
Sorbonne Université, CNRS, UMR 8235, Laboratoire Interfaces et Systèmes Électrochimiques, Paris, France

8.1 Introduction

The separation of the reactive and mechanical synergetic contributions to tribocorrosion damages, a quite challenging task, presented in this chapter is based on an original description of the electrochemical response of the interface to a tribological solicitation. Due to nonlinearity of fundamental tribological phenomena [1] and electrochemical laws (Volmer-Tafel, Butler-Volmer, Wagner-Traud, Stern-Geary, high field charge transfer, and higher-order reactions rate) [2, 3], large-amplitude transient regimes covering a wide range of mechanical and reactive conditions may be regarded as not the best-suited ones for investigating accurately the tribocorrosion phenomena at a quasi-steady state of mechanical and electrochemical constraints.

Two main cases are considered, namely applying a small amplitude sinewave modulation of angular velocity in a pin-on-disc configuration on one hand, and modulation of the electrochemical potential around a permanent regime on the other hand. In both cases, a transmittance, the frequency dependence of the response/perturbation ratio is derived, measured, and modeled based on a widely accepted class of models of passive film response to a mechanical and electrochemical disturbance. This approach takes advantage of a large body of instrumental and mathematical tools available for the harmonic analysis of physical systems, either in the time or frequency domain. Applied to stainless steel, the modulation of the sliding velocity provided new access to the repassivation transient in terms of time constants. On the other hand, modulation of the potential in the passive domain allowed separating the mechanical and reactive contributions to the global wear.

8.2 General concept, experimental approach, and techniques

A formally general case is considered in which a sliding pin removes all the surface passive film and further ploughs deeper into the underlying metal (Fig. 8.1). The quantity $|v| \cdot |F_t|$ is the mechanical power W_m injected in the system. It is consumed in irreversible processes of heat generation, plastic deformation, and breakdown of the metallic bounds [4] and transferred to reversible ones like elastic deformation. The right-hand side of the pin displays the initial surface of the material ahead of the pin which slides with the velocity v. The left-hand side shows schematically the film healing and the corresponding transient of the multicomponent current density $j(x)$, as a function of the distance $x = v \cdot t$ from the friction point. This function is simply transformed from the actual physical time dependence $j(t)$ through $x = v \cdot t$. This "traveling transient" attached to the pin is the basis of the derivation of the current response under modulation of the velocity, as the first example, or of the potential as the second example.

It must be emphasized that projecting $j(t)$, an intrinsic function resulting from pure reaction kinetics and issued from a separate model or experiment, over the space variable $x = v \cdot t$, is strictly rigorous only if the kinetic parameters (potential, temperature, and activities of species) are independent of x. The presence of ohmic potential drop and/or of composition or temperature gradients is therefore excluded from this representation. However, the charge balance applied locally to the partial electrochemical components of $j(x = v \cdot t)$ flowing through the worn track remains valid in all cases. The double-layer charging current j_{dl} is generally not mentioned in tribocorrosion and can be actually neglected [5] since the surface charge, around 1 m C cm^{-2}, being orders of magnitude smaller than that consumed in the electrochemical reactions associated with passivation, typically a fraction of C cm^{-2}.

Excepted in presence of a bulk oxidant active at this potential, the cathodic component j_c can be discarded, it flows essentially from the surrounding not abraded areas. It is vanishingly small under anodic polarisation.

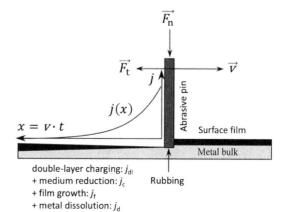

Fig. 8.1 Schematic time-space representation of abrading a film-protected metal by an electrochemically inert counter abrasive body and of the components of the resulting transient current density behind the counter-abrasive body.

If the number of electrons exchanged in the reaction is known, the anodic charge for film reconstitution having passed as j_f is a net measure of the quantity of metal lost by film wear. When the simultaneous metal dissolution j_d during the transient is added, $j(t)$ gives access to the total amount of metal lost by the so-called wear-accelerated corrosion but again only if the valence of dissolution is known. Of course, the metal removed directly by deep plowing (pure mechanical wear) is not included in this electrochemical determination. Normally, total loss equals the sum of wear accelerated corrosion plus mechanical wear. Although not emphasizing the synergism between chemical and mechanical degradation, this balance seems to have a certain mechanistic virtue [6].

Modeling the macroscopic response is made greatly easier at the expense of simplifying assumptions, some of them are often made implicitly in the literature:

- The surface ahead of the sliding point is supposed to be "virgin." In practice, this condition is physically unrealistic since it would require an endless sample in Fig. 8.1 in order to exclude any recursive sliding. However, this situation is fulfilled during the first stroke (reciprocating tribometers) or the first rotation (rotating pin-on-disc or disc-on-pin tribometers). Another possibility is equivalently a surface having recovered its initial full steady-state since the last passage of the sliding body. This would happen only at extremely low, and not physically significant, sliding velocities. It is expected, and often confirmed by the experiments, that despite a nonexplicitly formalized balance between surface sliding and film recovery, a periodic solicitation by reciprocating sliding ends in a periodic response and a continuous sliding in a steady-state one [3].
- The contribution to the current flowing from the body-body contacting area, supposed to be small and "dry," is generally not considered, except as a promising direction of research [4] and in an attempt to depict third-body effects taking place herein [7].

As proposed initially in [8], the current measured at any time is given by the integration, from the beginning of the experiment at $t = 0$, of the current density over the width w of the sliding track:

$$I(t) = w \cdot v \cdot \int_0^t j(t)dt \qquad (8.1)$$

where v is the modulus of $\vec{v}$. If the integral is convergent, it should be for expressing a permanent state of corrosion, $I(t)$ reaches a limiting value I_∞ given by Eq. (8.2) for t larger than the characteristic time of surface recovery.

$$I_\infty = w \cdot v \cdot Q \qquad (8.2)$$

where Q is the total surface density of charge flowing across the interface, in coulomb by unit area, over the whole transient. From a kinetics viewpoint, in Eq. (8.1) v plays a role identical to a rate constant in the expression of the electrochemical current. However, it must be pointed out that being an integral quantity, $I(t)$ does not give access directly to $j(t)$ and hence is not highly discriminative. Eq. (8.1) was exploited in various ways for interpreting many tribocorrosion measurements in transient regimes of sliding velocity and current [6]. It is known that even though a periodic current response may be recorded under some conditions, its interpretation is complicated by the time-dependence of the relative positions of the electrodes during the counter-body course [3]. These problems are overcome by the design of rotating tribometers, with a constant velocity sliding generally used. Elaborated intermittent regimes have also been introduced with this equipment for approaching fretting conditions and evaluating the synergetic aspects in tribocorrosion [9, 10]. In contrast, this chapter is dealing with small amplitude modulation around a constant velocity.

It is easily shown that, under the simplifying condition stated above, after one revolution the steady-state current is obtained by recasting Eq. (8.1) in the form:

$$I(t_{rot}) = \frac{w \cdot 2\pi \cdot R_{tr}}{t_{rot}} \cdot \int_0^{t_{rot}} j(t)dt \qquad (8.3)$$

where R_{tr} is the radius of the pin trajectory and t_{rot} the period of rotation.

According to this general model, a time-independent current does not mean at all that the electrode is in a true steady-state or a position-independent state. Compared with classical interfacial electrochemistry, the situation is totally exotic. In fact on the moving active area a point at a given fix position, at time "t" and distance $v \cdot t$ referred to the pin, is locked in a transient electrochemical state given by $j(t)$. The case is better depicted, as mentioned in Fig. 8.1, as a "traveling transient." This aspect is taken into consideration de facto in the derivation of steady-state and transient behaviors found in the literature. For example, in reciprocating tribometers, v is a piecewise function made of a sequence of constant velocity segments. $I(t)$ can still be computed by applying Eq. (8.1) successively to each segment, provided the initial conditions relative to each segment are stated. But the response of the current to a more general time program by which either the velocity, v, or the electrochemical potential, E, is varied in a continuous way, has to be derived. In the case of a small amplitude ac modulation of one of these driving forces, the model must be extended to account for the cumulated effect of the sine wave modulation on the local current density $j(t)$ at any value of the time t past the sliding point.

Fig. 8.2 displays the general equipment based on a pin-on-disc tribometer for measuring either the classical electrochemical impedance or the current/velocity transfer function, the so-called triboelectrochemical impedance [21]. The model of Fig. 8.1 is now applied to a circular trajectory resulting from the relative rotation of the sample vs. sliding pin. The equipment allows, with the same transfer function analyzer, to measure the electrochemical response of the sample under various combinations of potential-velocity [22,23]. The electrochemical conditions are set by the potentiostat while the mechanical ones, F_n and $v = \Omega \cdot 2\pi \cdot r$, are controlled by the tribometer interface. Experimental parameters and data acquisition are under PC control (not shown).

This equipment allows to handling the two transfer functions dealt with in this chapter. The first one is a mechano-electrochemical transfer function defined from the modulated component of the current $I(t)$ to a small amplitude sine wave modulation of the sliding velocity v. The second one is the classical electrochemical

Harmonic analysis of tribocorrosion

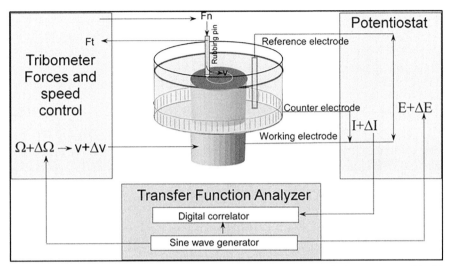

Fig. 8.2 General equipment for the electrochemical tribocorrosion techniques presented in this chapter. The tribometer parameters, rotation speed and forces, and the electrochemical parameters, potential, or current, are simultaneously controlled as steady-state and time-dependent quantities.

impedance spectroscopy (EIS) measured by a sine wave modulation of the applied potential. In this last case, contrasting with most of the literature impedance data in tribocorrosion obtained at the open-circuit potential E_{oc} [11, 12], the DC potential was fixed within the active-passive transition range in order to collect information mostly relative to the electrochemical conditions generally assumed to prevail in the abraded track.

8.3 Analysis of repassivation kinetics by ac modulation of the angular velocity of the counter-body

8.3.1 General formalism of the response to a time-dependent velocity of wear

This approach is based on the extension of Eq. (8.3) to a time dependent velocity of the sliding counter-body in a pin-on-disc contact. This is mathematically established as follows: at any time, the contribution to the total current $I(t)$ of a unit length section of the sliding track created at the time $(t - \tau)$, where τ is the time elapsed between the actual position, t, on the time axis and the time at which this element was generated. The velocity was then $v(t - \tau)$ and the current density is hence given by the product $j(\tau) \cdot v(t - \tau)$. In the general case, Eq. (8.1) is therefore recast as the convolution product of j and v:

$$I(t) = w \int_0^t j(\tau) \cdot v(t-\tau) d\tau \tag{8.4}$$

In order to define a linear response of $I(t)$, a small amplitude sine wave modulation $|\Delta v|e^{j\omega t}$, where ω is the angular frequency, is superimposed to the mean velocity. For a rotation period t_{rot} and an angular velocity $\Omega_0 = \frac{2\pi}{t_{rot}}$, the total current is expressed as the sum of the two components:

$$I(t_{rot}, t) = w \cdot R_{tr} \cdot \left(\Omega_0 \int_0^{t_{rot}} j(\tau) d\tau + \int_0^{t_{rot}} j(\tau) |\Delta \Omega| e^{j\omega(t-\tau)} d\tau \right) \tag{8.5}$$

The first term in the right hand side is the mean current response to Ω_0, identical to Eq. (8.3). The second one is the ac response of the current to the velocity modulation.

The potentiostatic triboelectrochemical admittance, TEA, is defined therefore as:

$$\frac{\Delta I}{\Delta \Omega} = w \cdot R_{tr} \left(\frac{\int_0^{t_{rot}} j(\tau) |\Delta \Omega| e^{j\omega(t-\tau)} d\tau}{|\Delta \Omega| e^{j\omega t}} \right) = w \cdot R_{tr} \int_0^{t_{rot}} j(\tau) e^{-j\omega \tau} d\tau \tag{8.6}$$

The TEA expressed by Eq. (8.6) is the one-sided ($t > 0$) Fourier transform of $j(t)$ in a time-window of length t_{rot}. As shown below, even though $j(t)$ is not readily extracted from the TEA, pertinent information on $j(t)$ can be gained by a simulation approach. It must be emphasized that Eq. (8.6) offers a totally original way for investigating the electrochemical response of a surface about a mean steady-state tribocorrosion regime while contributions are mostly confined to large transients between no abrasion and full abrasion conditions.

8.3.2 Experimental validation and application

All the results given below were recorded in the following electrochemical conditions: AISI 316L Stainless Steel, aqueous solution containing boric acid (1000 ppm of boron) and lithia (12 ppm of lithium). The electrochemical control of the working electrode submitted to sliding is classically performed by a Solartron 1286 potentiostat. The potential is maintained in the passive range at +0.1 V vs Ag/AgCl. The tribological conditions, namely normal force F_n and pin angular velocity Ω, are controlled by a P-o-D tribometer Falex and the associated RSAI software. The current flowing through the cell and the velocity of the pin are numerically stored using two channels of a PowerLab/8SP-ADInstruments. The working electrode is a disc of 25 mm in diameter and the circular sliding track has a radius $R_{tr} = 5$ mm. The counter-body is a zirconia rod, 7 mm in diameter, terminated by a spherical cap with a curvature radius, R_{cb} of 50 mm, the normal force F_n was set at 5 N. Fig. 8.3 displays a sample of the current trace under an ac modulation of Ω at 0.025 Hz. The response exhibits a reasonable sinewave shape but this result was further validated by Fourier analysis. The frequency spectrum shown in Fig. 8.4 indicates that the response contains a very low level of harmonics in spite of the large, 0.3, relative modulation ratio, hence validating the linearity conditions required by the impedance techniques.

Harmonic analysis of tribocorrosion

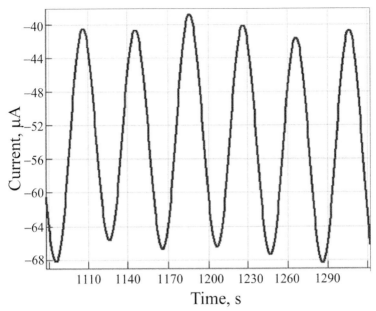

Fig. 8.3 Sample of the current response to the ac modulation of the rotation speed $\Omega_o = 5.2$ rad s^{-1}. Modulation frequency: $f = 0.025$ Hz, relative modulation ratio $\Delta\Omega/\Omega_o = 0.3$, $F_n = 5$ N. First-order low-pass filter, cut-off frequency 0.1 Hz.

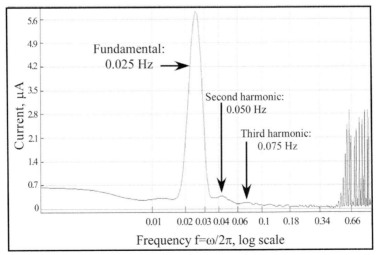

Fig. 8.4 Linear-log scale frequency spectrum of the current sample shown in Fig. 8.3. Second and third harmonics are very small. High frequency contribution is likely due to tribologic noises.

In view of the "exotic" character of TEA, basic instrumental and formal validations have to be performed. The consistency with the two other transfer functions defined at a same polarisation point [13], namely the electrochemical impedance Z and the triboelectrochemical-impedance: $TEI = \left(\dfrac{\Delta E}{\Delta \Omega}\right)_I$ is based on the relation (8.7):

$$\left(\dfrac{\Delta E}{\Delta I}\right)_\Omega = Z = -\dfrac{TEI}{TEA} \tag{8.7}$$

The ΔI and $\Delta \Omega$ data collected for frequencies between 0.01 and 10 Hz for a mean rotation speed of 20 rpm of the sliding pin can be processed frequency by frequency. As a result, the complex TEA diagram plotted in the Nyquist plane is shown in Fig. 8.5. The plot lies in the third quadrant of the complex plane (similarly to a capacitive electrochemical impedance plot, the imaginary axis is reversed as usual in EIS plots).

The low-frequency limit, positive, is in satisfactory agreement with the slope of the steady-state $I(\Omega)$ curve and the loop is consistent with a delayed rise of current following an increment of sliding velocity. The shape, clearly depressed, is not compatible with the exponential expression of the current density $j(t)$ which would generate a semicircle for the integral in Eq. (8.6) extending to infinity as expressed by Eq. (8.8).

$$\left(\dfrac{\Delta I}{\Delta \Omega}\right)_E = w \cdot R_{tr} \int_0^\infty e^{-\tau/\tau_1} e^{-j\omega\tau} d\tau = w \cdot R_{tr} \dfrac{\tau_1}{1+j\omega\tau_1} \tag{8.8}$$

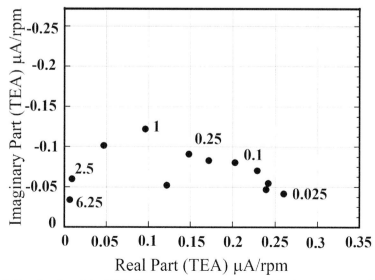

Fig. 8.5 Nyquist plot of the TEA. Same conditions as for Fig. 8.3. $\Omega_o = 2.1$ rad s^{-1}. Frequency labels in Hz.

Harmonic analysis of tribocorrosion

The same features were obtained at rotation speeds of 8 and 50 rpm with a frequency dispersion that is displayed clearly in the Bode plot (log of the modulus vs log of the frequency). Fig. 8.6 shows the experimental data points for three values of Ω_0 in the Bode modulus plane and the phase plane at 50 rpm. These results plots suggest rather than quite formal exploitation in terms of a CPE as usual in EIS, to propose an interpretation by the sum of two exponentials Eq. (8.9). In order to account for the windowing effect, the experimental data were hence compared with the numerical values generated by Eq. (8.6) for:

$$j(\tau) = \lambda_1 e^{-\tau/\tau_1} + \lambda_2 e^{-\tau/\tau_2} \qquad (8.9)$$

where λ_1 and λ_2 are the amplitudes and t_{rot} the mean rotation period corresponding to the actual Ω_o. The broken lines display the results of this simulation. The values of parameters yielding the best agreement are given in Table 8.1. They were estimated graphically by a trial-and-error procedure based on the modulus and phase Bode plots.

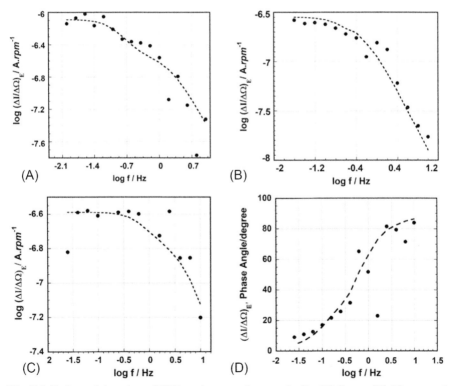

Fig. 8.6 Bode modulus plots of TEA at three rotation speeds Ω_o: (A) 8 rpm, (B) 20 rpm, and (C) 50 rpm. Same conditions as for Fig. 8.5. Bode phase plot (D) corresponding to data presented in (B). *Broken lines*: simulation with Eq. (8.6) with $j(\tau) = \lambda_1 e^{-\tau/\tau_1} + \lambda_2 e^{-\tau/\tau_2}$.

Table 8.1 Values of the parameters: preexponential factors, time-constants and surface charges determined by a simulation procedure of the TEA data for three rotation rates of the abrading pin shown in Figs. 8.5 and 8.6.

Ω_o (rpm)	$w\lambda_1$ (A m^{-1})	$w\lambda_2$ (A m^{-1})	τ_1 (s)	τ_2 (s)	λ_1 (A m^{-2})	λ_2 (A m^{-2})	$\tau_1\lambda_1$ (C m^{-2})	$\tau_2\lambda_2$ (C m^{-2})
8	0.000078	0.000468	1.5	0.1	0.156	0.936	0.234	0.0936
20	0.0000052	0.000156	3	0.3	0.0104	0.312	0.0312	0.094
50	0.00000338	0.000195	2	0.2	0.00676	0.39	0.0135	0.078

In the four right-hand side columns the parameters are rescaled by taking into account the actual track width, w, determined optically at the end of each experiments.

An estimate of the parameters is first made graphically and then manually refined to approach the best simulation. A quite satisfactory match is observed considering the rather large scattering of the TEA points. The main source of error is the parasitic modulation of the current at the rotation frequency due to a residual lack of levelness of the electrode surface with the pin trajectory, in spite of a careful machining.

In order to make these quantities more relevant to the actual phenomena, the amplitude λ_1 and λ_2 were rescaled in A m^{-1} by multiplication by w. The actual sliding track width w was measured on each particular disc at the end of the experiment. The value of w is dependent on the particular experiment considered, through the sliding conditions or the duration of the measurement.

It is clear that $j(\tau) = \lambda_1 e^{-\tau/\tau_1} + \lambda_2 e^{-\tau/\tau_2}$ depends on Ω_o, in contrast with the literature models which assume the same repassivation transient when changing the sliding rate or periodicity. The two time-constants τ_1 and τ_2 remained of the same order of magnitude and their ratio τ_1/τ_2 close to 10. The amplitude λ_1 decreased as the rotation speed goes up while λ_2 is a little dependent, with no clear trend. The consideration of the density of charge, integral of the current densities from $t = 0$ to t_{rot}, flowing under the two components of $j(t)$ brings about a more relevant insight since likely less sensitive to experimental dispersion. The results are given in the right-hand columns of Table 8.1, in surface density of charge. The charge $\tau_1\lambda_1$ associated with the shorter time-constant τ_1 can be regarded as constant while the charge $\tau_2\lambda_2$ having passed during the larger time constant τ_2 increases with the rotation period. One can infer that the time-constants correspond to two different surface processes. The shorter would reflect a fast process independent of the overall electrochemical activity, linked to Ω_o, of the surface, the longer one to a slower relaxation strongly coupled to the average degree of recovery of the surface. This is clearly not included in the model initially considered in this work and must be the starting point of an original effort of modeling in the field. Comparison with large current transients, recorded in the same triboelectrochemical conditions, after stepping to zero the pin velocity, can be found in [13].

8.4 From steady-state to electrochemical impedance ac modulation of the applied potential

The present section is aimed at a deeper insight into the application of electrochemical impedance spectroscopy (EIS) in triboelectrochemistry. Several works have been dealing with this technique for a better understanding of the cross-interaction between mechanical and reactive contribution to the total wear known as the synergetic effect. In most of them [11, 12] the technique is restrictively applied at the open-circuit-potential, Eoc, for extending to tribocorrosion the classical measurement of polarisation resistance, R_p, by an AC signal. More advanced and comprehensive application of EIS to tribocorrosion at several potentials in the cathodic, open-circuit, and passive domains must be mentioned [14, 15]. However, as for any other electrochemical measurement, except if local electrochemical measurements are performed, the impedance arises from the sum of the current densities flowing across the whole sample. A deeper understanding of EIS data in tribocorrosion of passive materials, where one must focus on the competition between anodic dissolution and repassivation of the worn area, requires absolutely a model of the local ac response as a function of the time t past the sliding contact.

8.4.1 General formalism of the current response to a time-dependent potential

The response of the transient current to a sine wave modulation of the potential was derived from in a way similar to the modulation of the sliding velocity dealt with in the Section 8.3. By converting time to space owing to a now constant velocity v. The total current flowing to the sliding track of length L is given by Eq. (8.1) recast in

$$I(L) = w \cdot \int_0^L j\left(\frac{x}{v}\right) dx \tag{8.10}$$

and returning to a time σ, referred to the sliding pin, while t is still for the running time in $E(t)$:

$$I(t_{rot}) = w \int_0^{L/v} v \cdot j(\sigma) d\sigma = w \cdot v \int_0^{t_{rot}} j(\sigma) d\sigma \tag{8.11}$$

The simplest form of $j(\sigma)$:

$$j(\sigma) = j_0 \exp\left(-\frac{\sigma}{\tau}\right) \tag{8.12}$$

is considered in this description, where τ is the time-constant, for two main reasons:
- it is a decreasing, integrable and easy to handle function,

it is most often involved in the formal modeling of physical systems being controlled by first order linear differential equations. This is the case in many descriptions of electrochemical interfaces by heterogeneous reactions coupled by surface intermediates, like dissolution and film growth. It must be pointed out that the treatment below is not strictly limited to an exponential response. It remains valid for any decreasing and integrable function tending to 0 at a reasonable t.

The same procedure developed for the DC current, Eq. (8.1), can be applied for deriving the electrical admittance of the whole track by summing now the local ac current densities induced by a sine wave modulation of the applied potential. This is illustrated in Fig. 8.7.

From Eq. (8.11) the AC potential modulation $|\Delta E|\exp j\omega t$ being applied at every point σ, the transient generates the local current density $j(\sigma) = a(E_0, \sigma, j\omega) \cdot |\Delta E|\exp(j\omega t)$, where $a(E_0, \sigma, j\omega)$ is the local electrochemical surface density of admittance, per time unit, of the interface at the DC polarisation E_0 and the same time position σ (or distance $x = v\ \sigma$ from the pin).

The ac component of the overall current flowing between $x = 0$ and $t_{rot} = L/v$ is therefore:

$$\Delta I(t_{rot}j\omega) = w \cdot v \cdot |\Delta E| \exp(j\omega t) \int_0^{t_{rot}} a(E_o, \sigma, j\omega) \cdot d\sigma \tag{8.13}$$

$a(E_0, \sigma, j\omega)$ is fixed by the local electrochemical transient state of the surface. In the next section, its expression will be derived in the framework of a dissolution-repassivation model. Therefore, like for a traditional electrode, the electrochemical behavior of the electrode under abrasion is made available through both its DC current and its impedance spectrum.

8.4.2 Reaction model of dissolution-repassivation

The kinetic model depicted in Fig. 8.8 is well-known and generally adopted for interpreting the anodic behavior of pure metals and alloys in media covering a broad range of aggressiveness. Steady-state current-voltage curves as well as impedance and RRDE behavior in a wide frequency domain, displaying inductances and/or negative

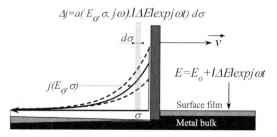

Fig. 8.7 Effect of a small amplitude ac modulation of the applied potential about E_o. The transient current oscillates between the two bounds in *dashed line*. The local current density Δj.

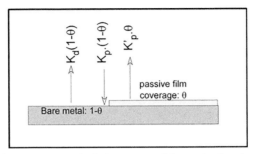

Fig. 8.8 Sketch of the components depicting a reversible passivation-depassivation and a dissolution of the bare area. At any instant, for a given E value, the state of the interface is totally determined by the value of the surface fraction θ occupied by the film.

resistance are explained by the relaxation of one (or several) surface coverage controlled by the mass balance between coupled reaction steps [16, 17].

- Classical case of a uniform, space independent, electrochemical state,
  The current density at steady-state, $\theta = \theta_s$ is given as:

$$j = F\left[n_p\left(K_p(1-\theta_s) - K_p'\theta_s\right) + n_d(K_d(1-\theta_s))\right] \quad (8.14)$$

where F is the Faraday constant, n_p and n_d the number of electrons exchanged in the reaction steps, K_p and K_p' the rate constants of respectively the passivation and depassivation reactions, θ_s the fractional coverage by the passive film. The detailed derivation of the impedance, Eq. (8.15), corresponding to this reaction scheme can be found in [18].

$$\frac{\Delta j}{\Delta E} = \frac{\Delta j_d}{\Delta E} + \frac{\Delta j_p}{\Delta E} = F\left[n_d K_d\left(b_d(1-\theta_s) - \frac{\Delta\theta}{\Delta E}\right) + n_p\beta j\omega\frac{\Delta\theta}{\Delta E}\right] \quad (8.15)$$

The reaction steps are assumed to obey Tafel laws, expressed in their condensed form: $K_p = k_p \exp(b_p E)$, $K_p' = k_p' \exp(b_p' E)$, $K_d = k_d \exp(b_d E)$, where b_p and $b_d > 0$ and $b_p' < 0$ are in V^{-1} related to the usual Tafel slopes B_i by $b_i = 2.303/B_i$, β is the surface concentration of the film material for $\theta_s = 1$, i.e., at full surface coverage, in moles cm^{-2}. One can see from Eq. (8.15) that the frequency dependence of Z arises only from the frequency response of the coverage θ, controlled by the mass balance:

$$\frac{\beta d\theta}{dt} = K_p(1-\theta) - K_p'\theta \quad (8.16)$$

- In the case of a tribocorrosion relevant situation, one has to consider a space-dependent electrochemical state. The DC and AC responses associated to the traveling transient in Fig. 8.7 are now expressed from the model shown in Fig. 8.8. The pin position is taken as the origin of the time σ and distance x scales. The derivation goes through the following steps:
  - Determination of $\theta(\sigma)$ by integration of Eq. (8.16) with $\theta(0) = 0$, a bare surface condition at the pin position in $\sigma = 0$,
  - Derivation of the transient current density $j(\sigma)$ by introducing $\theta(\sigma)$ in Eq. (8.14),

– Derivation of the admittance density at position σ by introducing the local values of $\theta(\sigma)$ instead of θ_s, and of $\frac{\Delta\theta}{\Delta E}(\sigma)$ instead of $\frac{\Delta\theta}{\Delta E}$ in Eq. (8.15).

Finally, the ac response of the whole abraded track is calculated from Eq. (8.13).

8.4.3 Application to the tribocorrosion of stainless steel in an acidic solution

The same stainless steel sample as in Section 8.3 above was investigated. A more aggressive medium, a 0.5 M H_2SO_4 solution, is selected in order to obtain higher corrosion of the depassivated metal. With the same goal and to better fulfill the validity conditions of the model, the sample was polarized, at 0 V vs Ag/AgCl at the onset of the passive domain (see Fig. 8.9). Tribological conditions are the same as in Section 8.3.

The quasi-steady state (1 m V s^{-1}) current-voltage curve of AISI 304L in Fig. 8.9 displays the typical profile for this material: successively by increasing potential, an activity peak, an active-passive transition with a negative slope, a passivity plateau, and a transpassive dissolution range. The domain of interest for the tribocorrosion behavior is the passivity range where AISI 304L is protected by a nanometer thin-oxide like film and its corrosion possibly resulting from both abrasive wear and reactive dissolution of the depassivated material. For a totally unpassivated surface, $\theta = 0$

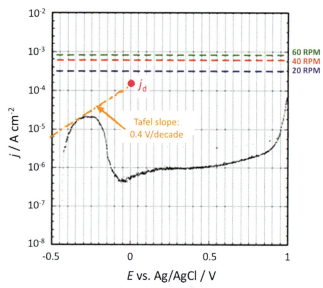

Fig. 8.9 Quasi-steady state (scanning rate: 1 m V s^{-1}) current-voltage curve of the SS sample in 0.5 M H_2SO_4. The *dot labelled* at 0.154 mA cm^{-2} corresponds to the dc current density, j_d, predicted by the model of Fig. 8.8 for a totally bare surface at $E = 0$ V/Ag/AgCl. The horizontal *dashed lines* indicate the upper limits of the current density $j_o = j(0)$ compatible with the condition $t_{rot}/\tau > 2.13$ for the three rotation rates considered.

Table 8.2 parameters values for the simulation of the steady-state and frequency response of the dissolution-passivation mechanism.

k_d (mol s^{-1} cm^{-2})	k_p (mol s^{-1} cm^{-2})	K_p' (mol s^{-1} cm^{-2})	b_d (V^{-1})	b_p (V^{-1})	b_p' (V^{-1})	β (mol cm^{-2})	n_d	n_p
8×10^{-10}	6.6×10^{-9}	5.4×10^{-12}	6	15	-15	1.3×10^{-9}	2	3

in Fig. 8.8, the set of kinetic constants shown in Table 8.2, optimized for simulating the dissolution-passivation model, generates the Tafel line with 400 mV/decade slope and a current density of dissolution of the bare metal j_d at 0 V/Ag/AgCl of 0.157 mA cm^{-2} as shown in Fig. 8.10.

These parameter values are reported on the experimental data plotted in Fig. 8.9. On this figure, from the evaluation of the charge involved in the transient regime, for every rotation speed applied, 20, 40, and 60 rpm, broken lines plotted indicate the upper limit of $j_o = j(0)$ allowing to match the validity condition $t_{rot}/\tau > 2.13$ for which most of the track is repassivated behind the moving pin. Above these values, the whole track remains almost in the active state and no effect of repassivation can be analyzed, even with this technique.

However, as shown in Fig. 8.1, the whole transient and its initial value at $\sigma = 0$, are associated not only with the dissolution rate but also with the film recovery. Therefore,

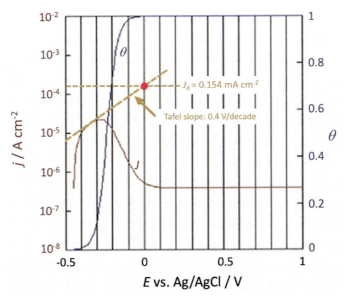

Fig. 8.10 Calculated potential dependence of the current density, left-hand scale and coverage θ, right-hand scale, with the parameters of Table 8.1. *Dashed line*, tangent to the $I(E)$ curve through the current density for $\theta = 0$ at $E = 0$ V.

the model can only be evaluated by considering the whole electrochemical response including the ac impedance. The impedance has been measured at $E = 0$ V/Ag/AgCl over the frequency range 62 kHz–0.016 Hz with seven frequencies per decade and an AC amplitude ΔE of 30 mV$_{rms}$. The raw data have been corrected of the contribution, measured separately, from the large passive area surrounding the sliding track. Fig. 8.11 shows the experimental Nyquist plot of the globally measured impedance and the impedance of the track alone. It is quite clear that any impedance-based work not including this correction may be largely irrelevant.

The impedance of the sliding track shows a short HF inflexion in the positive real quadrant, bends to the left, and then intersects the imaginary axis, clearly tending to a negative real limit at lower frequencies. This behavior in classical steady-state conditions is associated with a negative-sloped current-voltage curve, i.e., to passivation for pure metals [17, 19], and stainless steel [20]. This is intuitively consistent with the film recovering process that takes place behind the pin. However, more rigorous validation of the AC response of the current during tribocorrosion is provided only by comparing, to the simulated ones, the impedance diagrams at the three different rotation rates investigated. In order to visualize Eq. (8.13), the local admittances, $a(E_o, \sigma, j\omega)$ was computed along with the transient at a series of σ values between 0 and 3τ. Then the averaged value of the impedance was rescaled taking into account the track width w, from images and profilometry, and the velocity v, following Eq. (8.13).

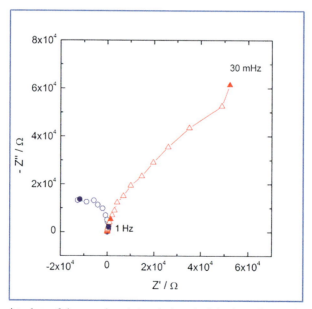

Fig. 8.11 Nyquist plots of the raw data *(triangles)*, and of the impedance of the track alone *(circles)*, after correcting for the contribution of the passive area in 0.5 M H$_2$S0$_4$ at $E = 0$ V/Ag/AgCl, $Fn = 5$ N, $\Omega = 40$ rpm.

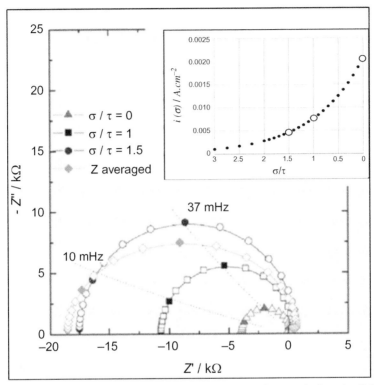

Fig. 8.12 Simulated local impedance densities at various σ values marked on the $j(\sigma)$ profile (insert) and impedance density averaged over the whole transient.

Fig. 8.12 shows some of the impedance plots calculated at several σ/τ values and the corresponding points on the $j(\sigma)$ curve in the insert. The averaged Z plot holds for the simulated response of the whole track. As expected, at short times a higher current density is associated with smaller impedance. The diagrams display a shape quite similar to the experimental one shown in Fig. 8.11 with a negative low-frequency limit on the real axis. Averaged computed impedances for the 3 rotation rates, once rescaled can be compared with the experimental diagrams as displayed in Fig. 8.13.

A fairly good agreement is found in terms of size, shape, and frequency location. The impedance is exactly inversely proportional to the abrasion velocity since, as highlighted by the model, the same $j(\sigma)$ transient is spread over an area proportional to the velocity. It should be stated that, in order to approach a correct simulation in the high-frequency range, essentially determined by the capacitive, dielectric-like response, not taken into account in the reactive model, an unusually large double-layer capacitance, of the order of 1 m F cm^2, had to be used, of the order of 1 m F cm^{-2}. That happened essentially for small σ values (that is for the beginning of the transient response) where the double layer is only starting to build-up and may show a larger capacitance. No data are available in the literature dealing with this aspect.

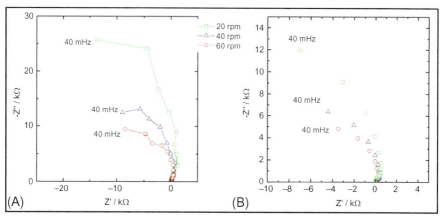

Fig. 8.13 Comparison of the experimental (A) and simulated (B) Nyquist plots of the whole track impedance at 3 rotation speeds in the experimental range of frequency.

8.4.4 Application to the separation of contributions to the total wear: Reactive and mechanical

The model is validated for both the steady-state and the AC responses under-wear, it yields a way of separating the contributions of the DC current to the film repair and the active dissolution of the fresh surface. This is definitively out of reach of any DC measurement which accesses only to the sum of these two reactive terms. As shown in Fig. 8.14 both components are very close to exponential decay in agreement with the first-order differential equation (Eq. (8.1)).

The ratio j_p/j_d is about 12 over the whole time range. It is clear that even in this aggressive media, the major part of the current under sliding is due to the film repair, not to the direct active dissolution of the unprotected surface. It can be considered that the reactive part of the total wear is traduced in j_p. This ratio of 12 may be regarded as a minimum since the β value in Table 8.1 is equivalent to a passive film only one to two monolayers thick. A larger β would of course lead to an even larger ratio. The total wear measured on the track profile [18] is a priori due to the sum of the reactive contribution, essentially j_p, and a likely pure mechanical wear of the underlying metal as represented in Fig. 8.1. This pure mechanical wear can be estimated by comparing the damaged profile to the faradic equivalent of $j_p + j_d$. A rough estimation by ascribing a dissolution valence of 2 to the removed material yields a charge of about 1C. This charge, taking into account the charge density flowing under the transient (381 µC cm^{-2}), the track area (0.077 cm^2) would require about 40,000 rotations. This figure is far larger than the cumulated number of rotations during the whole series of measurements, estimated to 10,000. It can thus be concluded that the pure mechanical damage can be estimated to the effect of 30,000 additional rotations, hence responsible of 75% for the total loss of material.

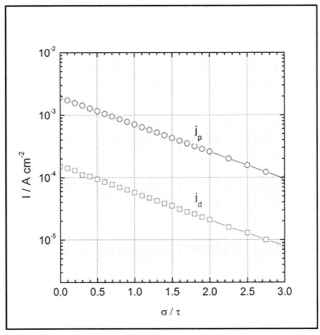

Fig. 8.14 Dissolution j_d and repassivation j_p components of the overall transient current densities as a function of the reduced time σ/τ.

8.5 Conclusions and perspectives

A new contribution to the understanding of the electrochemistry driven on a passive material under tribological conditions at a controlled potential is proposed. Small amplitude sine wave perturbations of two main driving forces in tribocorrosion, the slipping velocity or the applied potential are introduced. In both cases, the AC response of the sample is modeled by extending to a periodic regime of these parameters a widely applied model of tribocorrosion. Pin-on-disc equipment allowed to control the operational parameters while maintaining a constant cell geometry.

An original impedance function, the ratio of an electrochemical quantity over a mechanical one, is defined and its application illustrated. It provides the form of the repassivation kinetics attached to a given steady-state of friction, a piece of information generally considered as not accessible to the direct experiment. A two-time-constant response was evidenced and discussed in the framework of surface recovery and associated charge balance.

The electrochemical impedance of a surface undergoing friction was defined for the first time as the sum of the cumulative local contributions of the current density over the track. The model was applied to the behavior of stainless steel in the passivation range of potential. Impedance data under this unusual mode of a steady-state averaging space-dependent transients were validated and applied to the evaluation of the reactive and mechanical contribution to the total metal degradation.

8.6 Notations

b_i	Tafel coefficient (V^{-1})
B_i	Tafel slope (V)
E	potential (V)
E_{oc}	open circuit potential (V)
f	frequency (Hz)
F	Faraday's constant (C mol^{-1})
F_n	normal force applied by the tribometer (N)
I_∞	limiting value of the current (A)
$j(x)$	current density at the position x (A cm^{-2})
j_c	cathodic contribution to the current density (A cm^{-2})
j_d	metal dissolution contribution to the current density (A cm^{-2})
j_{dl}	current density for the double layer charging (A cm^{-2})
j_f	film growth contribution to the current density (A cm^{-2})
K_p	rate constant of the dissolution (mol s^{-1} cm^{-2})
K_p	forward rate constant of the passivation (mol s^{-1} cm^{-2})
K_p'	backward rate constant of the passivation (mol s^{-1} cm^{-2})
L	length of the sliding track (cm)
n_i	number of electrons exchanged
Q	total surface density of charge (C cm^{-2})
R_p	polarisation resistance (Ω cm^2)
R_{tr}	radius of the pin trajectory (cm)
t	time (s)
t_{rot}	period of rotation of the pin (s)
TEA	triboelectrochemical admittance (A rpm^{-1})
TEI	triboelectrochemical Impedance (V rpm^{-1})
v	velocity of the pin (rpm)
w	width of the sliding track (cm)
W_m	mechanical power (W)
x	position from the friction point (cm)
Z	electrochemical impedance (Ω cm^2)
β	maximum surface concentration (mol cm^{-2})
λ_i	preexponential factor (A cm^{-2})
σ	time (s)
θ	surface fraction occupied by the passive film
θ_s	surface fraction occupied by the passive film at steady-state
τ	time elapsed (s)
τ_i	time constant (s)
ω	angular frequency (rad s^{-1})
Ω	angular velocity of the tribometer (s^{-1})

References

[1] M. Urbakh, J. Klafter, D. Gourdon, J. Israelachvili, Nature 430 (2004) 525–528.
[2] S. Virtanen, in: D. Landolt, S. Mischler (Eds.), Tribocorrosion of Passive Metals and Coatings, Woodhead Publishing Lim, 2011, ISBN: 9780857093738 (Chapter 1).
[3] M. Keddam, F. Wenger, in: D. Landolt, S. Mischler (Eds.), Tribocorrosion of Passive Metals and Coatings, Woodhead Publishing Lim, 2011, ISBN: 9780857093738 (Chapter 8).
[4] P. Arnaud, S. Fouvry, S. Garcin, Wear 376–377 (2017) 1475–1488.
[5] P. Jemmely, S. Mischler, D. Landolt, Wear 237 (2000) 63–76.
[6] S. Mischler, Tribol. Int. 41 (2008) 573–583.
[7] D. Landolt, S. Mischler, M. Stemp, Electrochim. Acta 46 (2001) 3913–3929.
[8] S. Mischler, S. Debaud, D. Landolt, J. Electrochem. Soc. 145 (1998) 750–758.
[9] F. Wenger, M. Keddam, in: J.-P. Celis, P. Ponthiaux (Eds.), Testing Tribocorrosion of Passivating Materials Supporting Research and Industrial Innovation, European Federation of Corrosion Publications, 62, 119; Handbook, Published for the European Federation of Corrosion, 2012. ISBN: 1354-5116.
[10] N. Diomidis, in: J.-P. Celis, P. Ponthiaux (Eds.), Testing Tribocorrosion of Passivating Materials Supporting Research and Industrial Innovation, European Federation of Corrosion Publications, 62, 150, Handbook, Published for the European Federation of Corrosion, 2012. ISBN: 1354-5116.
[11] P. Ponthiaux, F. Wenger, D. Drees, J.-P. Celis, Wear 256 (2004) 459–468.
[12] V. Vignal, F. Wenger, B. Normand, in: J.-P. Celis, P. Ponthiaux (Eds.), Testing Tribocorrosion of Passivating Materials Supporting Research and Industrial Innovation, European Federation of Corrosion Publications, 62, 88, Handbook, Published for the European Federation of Corrosion, 2012. ISBN: 1354-5116.
[13] M.E. Orazem, B. Tribollet, Electrochemical Impedance Spectroscopy, John Wiley& Sons, Hoboken, NJ, 2011, p. 269.
[14] A.I. Muñoz, L.C. Julián, Electrochim. Acta 30 (2010) 5428–5439.
[15] J. Geringer, B. Normand, C. Alemany-Dumont, R. Diemiaszonek, Tribol. Int. 43 (2010) 1991–1999.
[16] M. Keddam, Anodic dissolution, in: P. Marcus (Ed.), Corrosion Mechanism in Theory and Practice, 2002, pp. 97–169, ISBN: 978-1420094626. New York, Basel.
[17] M. Keddam, W.R. Whitney award lecture: application of advanced electrochemical techniques and concepts to corrosion phenomena, Corrosion 62 (2006) 1056–1066.
[18] M. Keddam, F. Liao, P. Ponthiaux, V. Vivier, J. Solid State Electrochem. 19 (2015) 2591–2599.
[19] M. Keddam, O.R. Mattos, H. Takenouti, J. Electrochem. Soc. 2015 (1981) 266–274.
[20] M. Keddam, O.R. Mattos, H. Takenouti, Electrochim. Acta 31 (1986) 1159–1165.

Further reading

M. Keddam, P. Ponthiaux, V. Vivier, Electrochim. Acta 124 (2014) 3–8.
A. Berradja, D. Déforge, R.P. Nogueira, P. Ponthiaux, F. Wenger, J.-P. Celis, J. Phys. D Appl. Phys. 39 (2006) 3184–3192.
D. Déforge, F. Huet, R.P. Nogueira, P. Ponthiaux, F. Wenger, Corrosion 62 (2006) 514–521.

Modeling erosion-corrosion in metals: The effects of elastic rebound and impact angle on erosion-corrosion maps

M.M. Stack[a] and B.D. Jana[b]
[a]Department of Mechanical and Aerospace Engineering, University of Strathclyde, Glasgow, United Kingdom, [b]Department of Engineering and Mathematics, Sheffield Hallam University, Sheffield, United Kingdom

9.1 Introduction

Mathematical modeling of erosion-corrosion has been studied extensively in the literature [1–7] and one of the significant developments in the field has been the construction of erosion-corrosion maps, which predict the transition between erosion-corrosion regimes as a function of major influencing process parameters, such as erodent velocity and impact velocity.

A detailed methodology of constructing the erosion-corrosion maps is available in the literature [1–4], where it is evident that the generation of the maps requires predictive mathematical equations of the individual degradation process (i.e., erosion and corrosion). Therefore, the accuracy of the erosion-corrosion maps is vastly reliant on the underlying mathematical equations of the erosion and corrosion process.

Albeit predictive mathematical equations of solid particle erosion are widely available in the literature [8–14], the majority of these equations were derived for solid particle erosion, in which the erodent impact velocity is considerably higher than that might be expected in aqueous slurry erosion. The derivation of the predictive equations, therefore, assumed that erosion rate is proportional to the impact kinetic energy ignoring the elastic rebound effect of the erodent particles completely [1–14]. This assumption may not be reasonable for the slurry erosion conditions, in which the elastic forces are considerably high and therefore will cause the erodent particle to rebound from the impacting surface. An equation for the coefficient of restitution, e, will be subsequently derived in this chapter, which will prove that the extent of the rebound is inversely proportional to the impacting velocity. Therefore, the elastic rebound effect will be highly significant in aqueous slurry erosion.

Another important parameter that also influences the rate of erosion-corrosion is the impact angle at which an erodent particle arrives at the substrate. In modeling erosion-corrosion, the impact angle effect is often ignored despite the significant influence it has on the rate of erosion, which has been reported to show a peak at oblique impact angles [2, 3].

In the earlier publication of this handbook, erosion-corrosion maps for a range of materials have been presented, where the maps were constructed for dry gaseous erosion, neglecting the particle rebound effect completely [3]. In this chapter, the maps have been modified incorporating the effect of the particle rebound effect at low impact velocity and presented for a range of metals (i.e., Fe, Ni, Cu, and Al) at various erosion-corrosion conditions. Erosion-corrosion maps have also been presented for the above metals at a range of impact angles, pH, and electrochemical potentials. In addition, materials selection maps have also been presented at the modeling conditions studied.

9.2 Energy balance during a solid particle erosion event

In an erosion event, the kinetic energy of the impacting erodent particle is mostly consumed by the plastic deformation at the impact site, however, a considerable amount of the energy (between 1% and 10%) is lost in the form of particle rebound kinetic energy [2, 11]. An energy balance during a solid particle erosion event is presented in Table 9.1.

In modeling studies of solid particle erosion [13], it is commonly assumed that the initial kinetic energy of the impacting particle is equal to the work done in forming a permanent indentation on the metal substrate.

During a particle impact event, the metal substrate will exert a resistance force to the indenting erodent particle and the force can be given as

$$\text{Force} = H_s A \tag{9.1}$$

where H_s is the hardness of the metal substrate and A is the projected area of indentation for an erosion event.

In the impact event, if the particle penetrates a distance, dx, the work done can be given as

$$\text{Workdone} = H_s A \, dx \tag{9.2}$$

The total work done in forming the indentation can, therefore, be given as

$$H_s \int_0^d A \, dx = H_s V \tag{9.3}$$

where V is the volume and d is the depth of the indentation.

Table 9.1 Energy balance during a solid particle erosion event [2, 11].

Impact kinetic energy	100%
Dissipated in plastic work	90%
Elastic wave energy	1%–5%
Rebound kinetic energy	1%–10%

It is assumed that the kinetic energy of the impact is equal to the total work done, therefore,

$$\frac{1}{2}mU_i^2 = H_s V \quad (9.4)$$

where U_i is the incident velocity of the impacting particle.

However, if the particle rebound effect is not ignored, the total work done should be equal to the difference in the kinetic energy of the impacting particle and the kinetic energy of the rebounding particle. Thus, the following equation is valid when the particle rebound effect is considered.

$$\frac{1}{2}mU_i^2 - \frac{1}{2}mU_r^2 = H_s V_r \quad (9.5)$$

where U_r is the particle rebound velocity and V_r is the relaxed (or permanent) indentation volume.

To solve Eq. (9.5) for the rebound velocity, it is necessary to derive an expression for the coefficient of restitution, e, which, at normal impact erosion, is a ratio of the rebound velocity to the incident velocity of the erodent particle. Whereas, at an oblique impact erosion, the expression considers the normal components of both the incident and rebound velocities.

In this analysis, it is assumed that a spherical particle impacts and rebounds in the perpendicular direction to the target surface. Therefore, the coefficient of restitution can be defined as:

$$e = \frac{U_r}{U_i} \quad (9.6)$$

9.3 Derivation of an expression for the coefficient of restitution

During an erodent impact event, if it is considered that a significant proportion of the stored elastic energy is released, the indentation (or crater) formed on the target metal will initially be in its unrelaxed state and subsequently following the release of the elastic stored energy will become in its relaxed state. The release of stored elastic energy will cause the particle to rebound from the target metal surface. If the particle rebound effect is considered negligible, the total kinetic energy of the impacting particle can be equated to the work done in forming the unrelaxed crater.

$$\frac{1}{2}mU_i^2 = H_s \cdot V_{ur} \quad (9.7)$$

where m is the mass of the erodent particle, respectively, and V_{ur} is the volume of an unrelaxed crater produced by the particle impact on the metal target.

The volume of the unrelaxed crater, V_{ur}, can be expressed as

$$V_{ur} = \frac{\pi w^4}{64r} \tag{9.8}$$

where r is the radius of the impacting particle and w is the diameter of the unrelaxed crater (which is assumed to be spherical). In deriving Eq. (9.8) it was assumed that the radius of the impacting particle is significantly greater than the depth of the unrelaxed crater. Now, combining Eqs. (9.7), (9.8), the following relationship can be obtained

$$\frac{1}{2}mU_i^2 = H_s \cdot \frac{\pi w^4}{64r} \tag{9.9}$$

Assuming $m = 4/3\, \pi r^3 D_p$, where D_p is the density of the erodent particle, an expression for the unrelaxed crater diameter can be derived from Eq. (9.9)

$$\frac{w}{r} = 2.56 U_i^{0.5} \left(\frac{D_p}{H_s}\right)^{0.25} \tag{9.10}$$

A similar expression was derived by Tirupataiah et al. [14].

It should be noted that Eq. (9.10) is valid only when the rebound velocity is negligibly small compared with the incident velocity, which means the coefficient of restitution is close to zero.

However, when the rebound velocity is not negligible, especially at lower incident velocities, the relaxation of the crater occurs. Tirupataiah et al. [14] and Tabor [15] suggested that the relaxation takes place mainly in the direction of the crater depth and the diameter of the crater remains unaffected in the relaxation process. Therefore, the volume of the relaxed crater can similarly be expressed.

Tirupataiah et al. [14] suggested that the momentarily stored elastic energy, following the particle impact, is subsequently released causing the particle to rebound from the impact site. Therefore

$$\text{Stored elastic energy} = \frac{1}{2}mU_r^2 \tag{9.11}$$

In their analysis, Tirupataiah et al. [14] used the following expression for the stored elastic energy derived by Hertz:

$$\text{Stored elastic energy} = 0.7 \left(\frac{H_s^2}{E_e}\right) w^3 \tag{9.12}$$

E_e in Eq. (9.12) is defined as follows:

$$E_e = \frac{E_b E_t}{E_b(1-v_t^2) + E_t(1-v_b^2)} \tag{9.13}$$

where E is the elastic modulus υ is the Poison's ratio, the subscript b and t refer to the ball/particle and target, respectively.

Combining Eqs. (9.7) and (9.10)–(9.12), the final expression for the coefficient of restitution can be given as:

$$\frac{1}{2}mU_r^2 = 0.7\left(\frac{H_s^2}{E_e}\right)w^3 \tag{9.14}$$

$$\frac{2}{3}\pi r^3 D_p U_i^2 e^2 = 0.7\left(\frac{H_s^2}{E_e}\right)w^3 \tag{9.15}$$

$$e = 0.578\left(\frac{H_s^2}{E_e D_p U_i^2}\right)^{\frac{1}{2}}\left(\frac{w}{r}\right)^{\frac{3}{2}} \tag{9.16}$$

Substitution of Eq. (9.10) into Eq. (9.16) gives

$$e = 2.37\frac{H_s^{0.625}}{E_e^{0.5}D_p^{0.125}U_i^{0.25}} \tag{9.17}$$

The expression for the coefficient of restitution, Eq. (9.17), was based on the analysis used by Tirupataiah et al. [14]. Several identical expressions to Eq. (9.17), except the numerical constant, were reported by various other researchers in the field, and their reported numerical constants are shown in Table 9.2.

9.4 Derivation of erosion models based on the rebound analysis

As stated earlier, the expression for the coefficient of restitution, Eq. (9.17), was derived for normal impact erosion events, where the impact and rebound of the particle are normal to the target surface. However, for simplicity, Eq. (9.17) was also assumed to be valid for the oblique impact erosion model derived in this work.

Table 9.2 The numerical constants for Eq. (9.17), reported in the literature.

Author	Numerical constant
Eq. (9.17)	2.37
Stack, Corlett, and Zhou [2]	1.36
Singer and Evans [16]	1.47
Tabor [15]	1.72
Johnson [17]	1.72
Brenner et al. [18]	2.73

9.4.1 The effect of particle rebound on the erosion models

In deriving Eq. (9.17), it was assumed that the stored elastic energy is released in the form of the kinetic energy of the particle rebound. Therefore, the energy required for plastic deformation of the target (or deformation energy) is equal to the difference in the kinetic energy of the impacting particle and that of the rebounding particle. Thus,

$$\text{Deformation energy} = \frac{1}{2}mU_i^2 - \frac{1}{2}mU_r^2 \tag{9.18}$$

The deformation energy in Eq. (9.18) may be equated to the kinetic energy associating the deformation velocity (U_d), at which the target surface is plastically deformed by the particle impact.

$$\frac{1}{2}mU_d^2 = \frac{1}{2}mU_i^2 - \frac{1}{2}mU_r^2 \tag{9.19}$$

As $e = U_r/U_i$, thus Eq. (9.19) becomes:

$$U_d = U_i\left(1 - e^2\right)^{0.5} \tag{9.20}$$

To derive erosion models based on the particle rebound effect, it is assumed that the dimensionless erosion is proportional to the deformation velocity rather than the impact velocity of the erodent particles. Therefore, using Eq. (9.20), the erosion rate (K_{Ereb}) may be derived from the dimensionless erosion (E_{reb}) expressions for both normal and oblique impact erosions already existing in the literature.

9.4.2 Dimensionless erosion based on erodent rebound effect

In the literature, the commonly used term for predicting erosion rate is a dimensionless number (E), which is expressed as the mass of material removal per unit mass of erodent particles.

For normal impact erosion, the dimensionless erosion derived by Sundararajan and Shewmon [10] was modified based on the particle rebound effect presented in this article. Details of their model, given erosion-corrosion modeling, have been reported earlier [3]. The expression for the dimensionless erosion derived by Sundararajan and Shewmon [10] may be presented as follows:

$$E = \frac{6.5 \times 10^{-3} U_i^{2.5} D_p^{0.25}}{C_p T_m^{0.75} H_s^{0.25}} \tag{9.21}$$

where U_i is the erodent impacting velocity, T_m and C_p are the melting point and specific heat capacity of target metal, respectively. D_P and H_s are the density of erodent particles and static hardness of target material, respectively.

Modeling erosion-corrosion in metals

Combining Eqs. (9.20), (9.21), the final expression of the dimensionless erosion rate based on particle rebound (E_{reb}) at normal impact can be given as follows:

$$E_{reb} = \frac{6.5 \times 10^{-3} D_p^{0.25} U_i^{2.5} (1-e^2)^{1.25}}{C_p T_m^{0.75} H_s^{0.25}} \tag{9.22}$$

For oblique impact erosion, the dimensionless erosion (E) derived by Finnie [19] (commonly known as Finnie's second model) is modified based on the particle rebound effect. Details of Finnie's model can also be found in [3]. The dimensionless erosion expression derived Finnie may be presented as follows:

$$E = \left(\frac{f_c}{2k}\right) \frac{D_t U_i^2}{0.9272 H_S} \left(\sin 2\alpha - \frac{8}{k}\sin^2\alpha\right) \text{ for } \tan\alpha \leq \frac{k}{8} \tag{9.23}$$

$$E = \left(\frac{f_c}{16}\right) \frac{D_t U_i^2}{0.9272 H_S} (\cos^2\alpha) \text{ for } \tan\alpha \geq \frac{k}{8} \tag{9.24}$$

where f_c is the proportion of particles impacting the surface cutting in an idealised manner, which is 10%, as suggested by Finnie. α is the particle impact angle. Other parameters of these equations are defined in Appendix-I, Nomenclature.

Combining Eqs. (9.20), (9.23), (9.24), the final expression of the dimensionless erosion rate based on particle rebound (E_{reb}) at oblique impact can be given as follows:

$$E_{reb} = \left(\frac{f_c}{2k}\right) \frac{D_t U_i^2 (1-e^2)}{0.9272 H_S} \left(\sin 2\alpha - \frac{8}{k}\sin^2\alpha\right) \text{ for } \tan\alpha \leq \frac{k}{8} \tag{9.25}$$

$$E_{reb} = \left(\frac{f_c}{16}\right) \frac{D_t U_i^2 (1-e^2)}{0.9272 H_S} (\cos^2\alpha) \text{ for } \tan\alpha \geq \frac{k}{8} \tag{9.26}$$

9.4.3 Derivation of pure erosion rate from the dimensionless erosion

In this study, the rate of erosion is expressed as the mass of material removed per unit area of eroded surface per unit time (or in kg m^{-2} s^{-1}), and this unit is used to express the loss of material under both the erosion and corrosion conditions studied.

To derive an expression for the rate of erosion rate (in kg m^{-2} s^{-1}), the dimensionless erosion rate is required to be multiplied by the particle flux, which is the total mass of particle impacting the target surface per unit area per unit time. The expression for particle flux is as follows:

$$\text{Particle flux} = \frac{\text{Mass of impacting particle}}{\text{Unit area} \cdot \text{Unit time}} = c \cdot U_i \tag{9.27}$$

where c is the particle concentration in the slurry (kg m^{-3}) and U_i is the particle impact velocity (m s^{-1}).

In an erosion event, it may not be the case that all the particles present in the slurry impact the target surface. Therefore, Eq. (9.27) is modified by considering a fraction (f_i) of particle concentration responsible for the erosion loss. Therefore, Eq. (9.27) becomes

$$\text{Particle flux} = \frac{\text{Mass of impacting particle}}{\text{Unit area} \cdot \text{Unit time}} = f_i c \cdot U_i \tag{9.28}$$

Hence, the final expression for the erosion rate (K_E) can be derived as follows:

$$K_E = E \cdot f_i \cdot c \cdot U_i \tag{9.29}$$

The expression of the particle concentration can be given as

$$c = \frac{sol \cdot D_p \cdot D_w}{sol \cdot D_w + D_p(100 - sol)} \tag{9.30}$$

where "sol" is the percent solid by mass. D_p and D_w are the density of the particle and water, respectively.

Combining Eqs. (9.22), (9.29), the total erosion rate at normal impact (K_{Ereb}) can be given as:

$$K_{Ereb} = \frac{6.5 \times 10^{-3} D_p^{0.25} f_i c U_i^{3.5} (1-e^2)^{1.25}}{C_P T_M^{0.75} H_S^{0.25}} \tag{9.31}$$

Similarly, combining Eqs. (9.25), (9.26), (9.29), the total erosion rate at oblique impact angle (K_{Ereb}) can be given as:

$$K_{Ereb} = \left(\frac{f_c}{2k}\right) \frac{D_f_i c U_i^3 (1-e^2)}{0.9272 H_S} \left(\sin 2\alpha - \frac{8}{k}\sin^2\alpha\right) \text{ for } \tan\alpha \leq \frac{k}{8} \tag{9.32}$$

$$K_{Ereb} = \left(\frac{f_c}{16}\right) \frac{D_f_i c U_i^3 (1-e^2)}{0.9272 H_S} (\cos^2\alpha) \text{ for } \tan\alpha \geq \frac{k}{8} \tag{9.33}$$

9.5 Corrosion rate and boundary conditions

9.5.1 Corrosion rates at active and passive conditions

The derivation of the corrosion models for the construction of erosion-corrosion maps are given in [1–4]. For ease of understanding the final expression for the corrosion rate, relevant to this work, is given here. The final expression for the rate of corrosion (K_c) under active conditions can be given as

$$K_C = \frac{RAM \cdot i_a}{n \cdot F} \qquad (9.34)$$

where RAM is the relative atomic mass of target materials, i_a is the anodic current density, n is the number of electrons donated by the dissolution reaction, and F is the Faradays constant, equal to 96,485 C mol^{-1}.

The corrosion rate under passivation conditions can be given as

$$\Delta K_C = \frac{\text{const.} h D_f f_i c U_i^2}{r D_p^{0.5} H_S^{0.5}} \qquad (9.35)$$

where ΔK_c is the additive effect of erosion affected corrosion, D_f is the density of the passive oxide film, h is the thickness of passive film form on the metal surface, and r is the radius of erodent particle. The value of "const." for various metals (dimensionless) are given in Table 9.3.

9.5.2 Boundary conditions for the erosion-corrosion maps

In this work, erosion-corrosion maps are constructed in two different forms, namely regime maps and wastage maps. The regime maps identify the inherent mechanism of material loss, whereas the wastage maps identify the rate of that as high, medium, and low.

Boundary conditions used for the construction of the erosion-corrosion maps are stated in this section.

The total rate of erosion-corrosion (K_{EC}) can be expressed as

$$K_{EC} = K_{CO} + \Delta K_C + K_{EO} + \Delta K_E \qquad (9.36)$$

where K_{CO} is pure corrosion rate in the absence of erosion, K_{EO} is pure erosion rate in the absence of corrosion, ΔK_c is the effect of erosion on corrosion (additive effect), ΔK_E is the effect of corrosion on erosion (synergistic effect).

Eq. (9.36) can be simplified as

$$K_{EC} = K_C + K_E \qquad (9.37)$$

Table 9.3 Values of "const." for different metals in Eq. (9.35).

Metal	Fe	Ni	Cu	Al
Constant in Eq. (9.35)	86	96.7	98.3	65.1

where

$$K_C = K_{CO} + \Delta K_C \qquad (9.38)$$

and

$$K_E = K_{EO} + \Delta K_E \qquad (9.39)$$

For the construction of regime maps, the boundaries of the regimes are based on the ratio of the total corrosion rate (K_C) to the total erosion rate (K_E), as is given below:

$$\frac{K_C}{K_E} = \frac{K_{CO} + \Delta K_C}{K_{EO} + \Delta K_E} \qquad (9.40)$$

The erosion-corrosion regimes were defined according to the following boundary conditions. Please note the term dissolution or passivation is used instead of corrosion to identify the specific mechanism of corrosion.

Erosion dominated:	$K_C/K_E < 0.1$
Erosion-dissolution:	$0.1 \leq K_C/K_E < 1$
Dissolution-erosion:	$1 \leq K_C/K_E < 10$
Dissolution dominated:	$K_C/K_E \geq 10$

For the construction of wastage maps, the boundaries separating the regions of wastage (i.e., high, medium, and low wastage) are based on the total erosion-corrosion rate (K_{EC}) in mm y^{-1}.

Low:	$K_{EC} < 1$
Medium:	$1 \leq K_{EC} < 10$
High:	$K_{EC} \geq 10$

9.6 Elastic rebound effects on the transition boundaries of the aqueous erosion-corrosion map

9.6.1 Velocity—Potential maps for a range of pure metals

In order to understand the effect of erodent impact angle on the rate of erosion for the range of pure metals studied (i.e., Fe, Ni, Cu, and Al), the "dry" solid particle erosion rate as a function of impact angle is plotted in Fig. 9.1, in which Finnie's second model (Eqs. 9.32, 9.33) [8, 9, 19], and Sundararajan and Showman model (Eq. 9.31) [10] are used for the calculation of "dry" solid particle erosion at oblique impact (0–30 degrees) and normal impact, respectively. It should be noted that the data is extrapolated between the impact angle 30 and 90 degrees. As the pure metals studied are all ductile metals, Fig. 9.1, as expected, shows typical ductile erosion behavior of

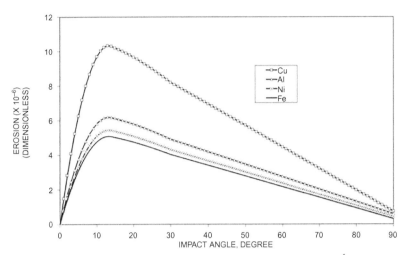

Fig. 9.1 Dimensionless erosion vs impact angle graph (for velocity 10 m s^{-1})—extrapolating from Finnie's second model [8, 16] for 0–30 degrees to Sundararajan and Shewmon's model [9] at 90 degrees.

the materials, with a peak around 13 degrees impact angle for all the metals studied. It is also evident in Fig. 9.1 that the erosion rate of Cu is the highest and Fe is the lowest along all the pure metals studied.

To understand the change in corrosion regimes of these metals, their Pourbaix diagrams are important for this study, and therefore they are shown in Fig. 9.2A–D. Please note here that the potential in the Pourbaix diagram is expressed against the standard hydrogen electrode while the "applied" potential in the erosion-corrosion maps is expressed against the standard calomel electrode.

As evident in Fig. 9.2, both Fe and Ni actively dissolve at low potentials and passivate at high potentials; the dissolution and passivation regimes for Ni extend to higher potentials compared with Fe. Cu actively dissolves at low pHs and passivated at high pHs; it also shows a larger immunity region compared to the other metals, but Al passivates at intermediate pHs and actively dissolves at both low and high pHs.

Regime and wastage maps for both 10 and 90 degrees impact angles at pH 5 are shown in Figs. 9.3–9.6. Please note here that the "applied potential" is synonymous with "potential." The methodology for the construction of maps is given elsewhere [1, 3, 4]. In Figs. 9.5 and 9.6 different shadings are used to highlight the "velocity vs. potential zones" of the low, medium, and high erosion-corrosion rates (KEC) as defined at the end of Section 9.5.2.

The maps for 10 degrees impact are largely similar to those maps for 90 degrees impact, except the regime transition in the oblique impact maps appears at a considerably lower impact velocity compared to that in normal impact maps. This is because the erosion rate at oblique impact is significantly higher than that at normal impact, as shown in Fig. 9.1.

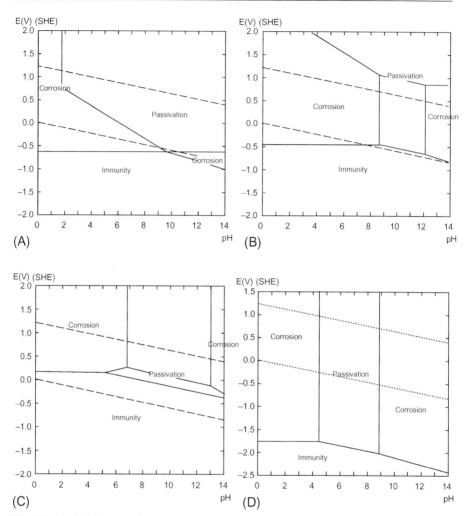

Fig. 9.2 Pourbaix diagram for (A) Fe in H_2O assuming passivation by formation of Fe_2O_3, (B) Ni in H_2O assuming passivation by formation of NiO, (C) Cu in H_2O assuming passivation by formation of CuO, and (D) Al in H_2O assuming passivation by formation of Al_2O_3 [20].

Modeling erosion-corrosion in metals

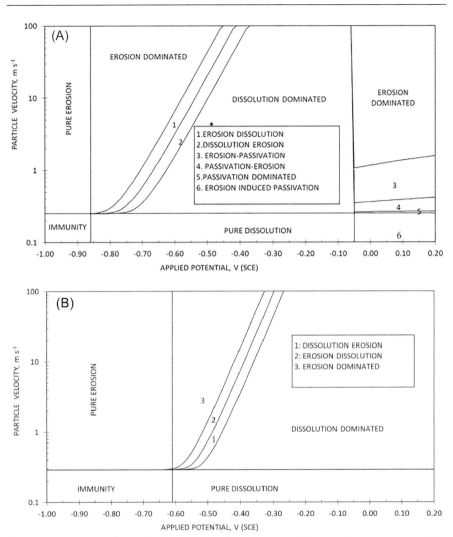

Fig. 9.3 Particle velocity-applied potential regime maps for 10 degrees impact angle at pH 5: (A) Fe, (B) Ni,

Continued

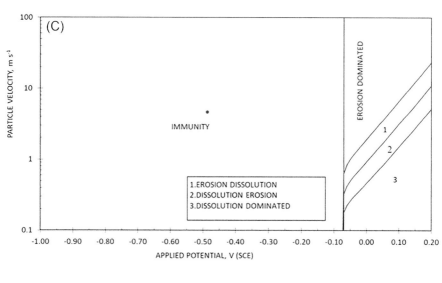

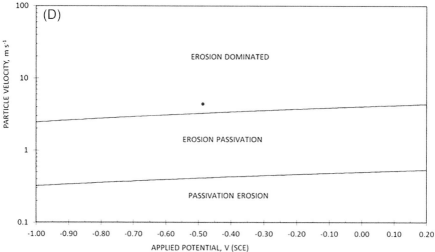

Fig. 9.3, cont'd (C) Cu, (D) Al.

Modeling erosion-corrosion in metals

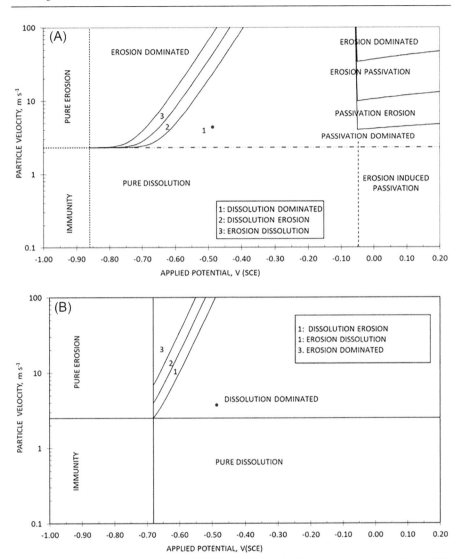

Fig. 9.4 Particle velocity-applied potential regime maps for 90 degrees impact angle at pH 5: (A) Fe, (B) Ni,

Continued

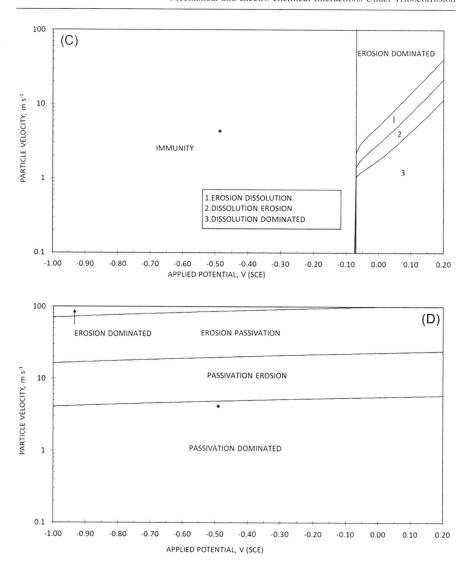

Fig. 9.4, cont'd (C) Cu, (D) Al.

Modeling erosion-corrosion in metals 207

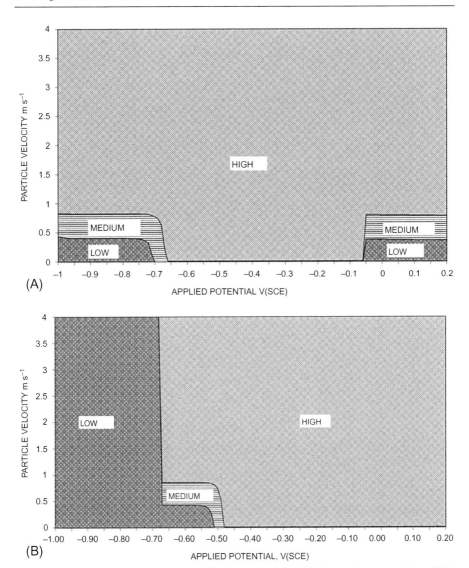

Fig. 9.5 Particle velocity-applied potential wastage maps for 10 degrees impact angle at pH 5: (A) Fe, (B) Ni,

Continued

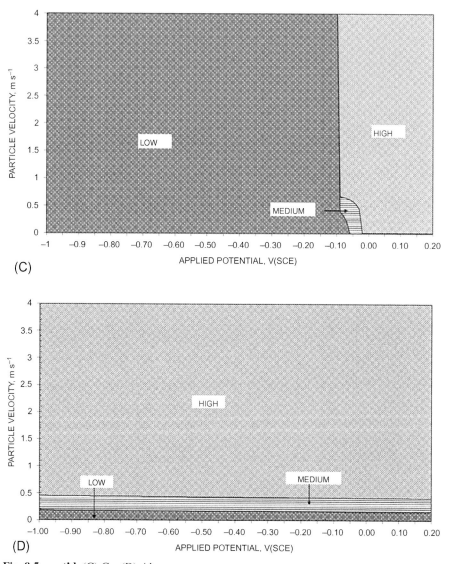

Fig. 9.5, cont'd (C) Cu, (D) Al.

Modeling erosion-corrosion in metals

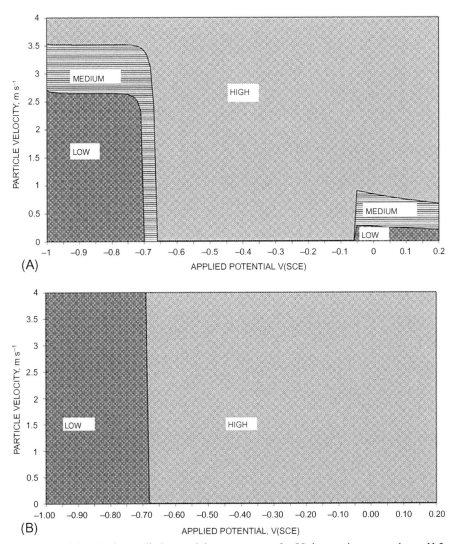

Fig. 9.6 Particle velocity-applied potential wastage maps for 90 degrees impact angle at pH 5: (A) Fe, (B) Ni,

Continued

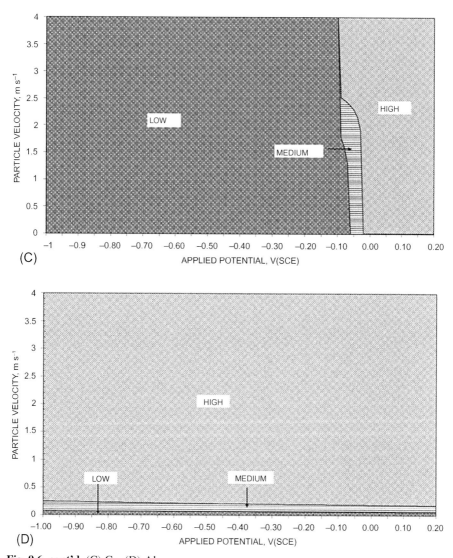

Fig. 9.6, cont'd (C) Cu, (D) Al.

Particle velocity-potential regime maps for pH 5 at 10 and 90 degrees, Figs. 9.3 and 9.4, show that Fe, Ni, and Cu actively corrode in the condition modeled. Cu, as expected, shows a large area of immunity compared with the other metals. Al is, however, fully passive in the condition modeled and showed passivation assisted erosion-corrosion regimes in the condition modeled. A comparison of the oblique and normal impact maps reveals that the threshold velocity for erosion at oblique impacts is comparatively lower than that at normal impacts. Additionally, it can also be observed that the transition velocity for the erosion-corrosion regimes is lowered in these maps compared to that in the maps for normal impact.

The erosion-corrosion wastage maps for pH 5 at 10 and 90 degrees, Figs. 9.5 and 9.6, show the change of wastage (i.e., high, medium, and low contours) of the individual metal with the applied potential. The wastage maps of Fe and Ni are broadly similar and the high wastage for these metals is consistent with dissolution affecting the erosion-corrosion process observed in their regime maps. Among all the metals studied, Cu shows the largest area of low wastage zone; this is consistent with the pure erosion regime (due to immunity) observed in the regime map of the metal. The map for Al shows a predominantly high wastage region albeit the metal being passive in the condition modeled; this is mainly because of the poor mechanical properties leading to preferential removal of the oxide film and the less adherence nature of the passive film on the metal surface. Although the oblique impact maps are qualitatively similar to the maps at normal impact, close scrutiny reveals that the transition velocity of the erosion-corrosion wastage regime is lower in the case of an oblique impact compared to that of the normal impact. In other words, all these metals possess a lower resistance to the erosion corrosion-corrosion process at 10 degrees impact compared to that at normal impact.

Materials performance maps, Fig. 9.7A and B, are constructed by superimposing the low waste region from the wastage map of the individual metals, Figs. 9.5 and 9.6. These materials performance maps can be used for material selection in aqueous slurry erosion applications; however, these maps need to be extended to the erosion-corrosion conditions of the applications (if those fall outside the conditions modeled here). Please note here that, although these maps were constructed by superimposing the "low waste" region of the individual metals, the "medium + high wastage" region was shown in the maps to indicate the modeling conditions at which all the metals become unsuitable.

As evident in Fig. 9.7, the resistance to erosion-corrosion of the individual metal is significantly different. Among all the metals studied, Cu shows the highest resistance to erosion-corrosion, and Al shows the lowest in the condition modeled. Fe shows resistance to the degradation process up to a significant impact velocity at both low and high potentials. A comparison of the material performance maps at 10 degrees with that at 90 degrees impact, reveals that the application of Fe is preferred at lower velocities in the case of 10 degrees impact. Surprisingly, the Al at 10 degrees impact showed resistance to the process up to a marginally higher velocity, compared to that at normal impact. Possible reasons for this response of Al are not yet clear.

9.6.2 Velocity—Impact angle map for a range of pure metals

Velocity-impact angle maps are presented in Figs. 9.8–9.12. Like the velocity-potential maps (Figs. 9.3–9.6); these maps are also presented in two different forms, namely regime and wastage maps and are constructed for both pH 5 and 9 at a constant potential of -0.45 V.

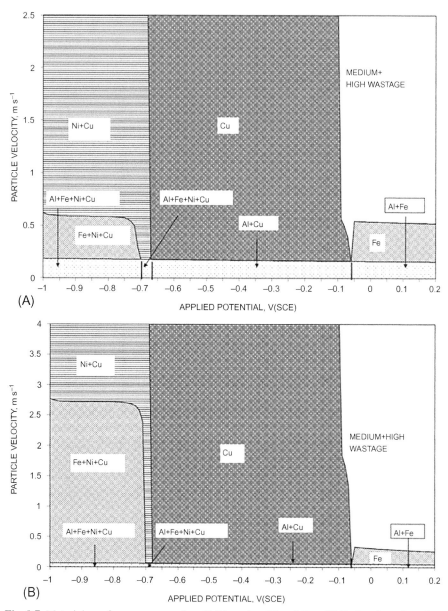

Fig. 9.7 Materials performance maps for pH 5 based on Figs. 9.5 and 9.6, showing where low wastage is observed for the various pure metals: (A) 10 degrees impact angle and (B) 90 degrees impact angle.

Modeling erosion-corrosion in metals 213

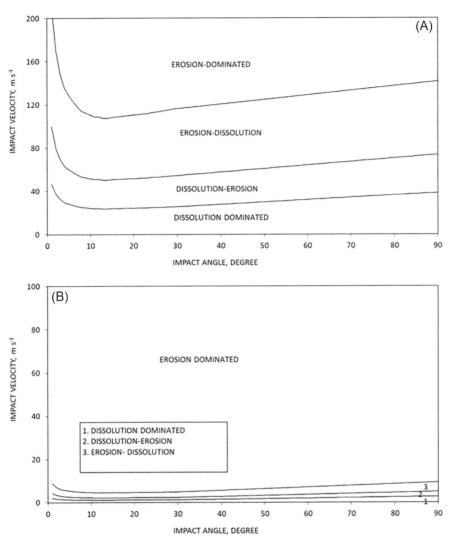

Fig. 9.8 Particle impact velocity-impact angle regime maps for pH 5 at -0.45 V: (A) Fe, (B) Ni,

Continued

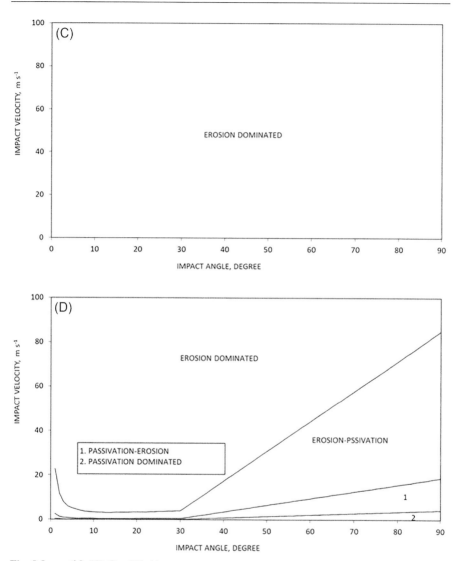

Fig. 9.8, cont'd (C) Cu, (D) Al.

Modeling erosion-corrosion in metals

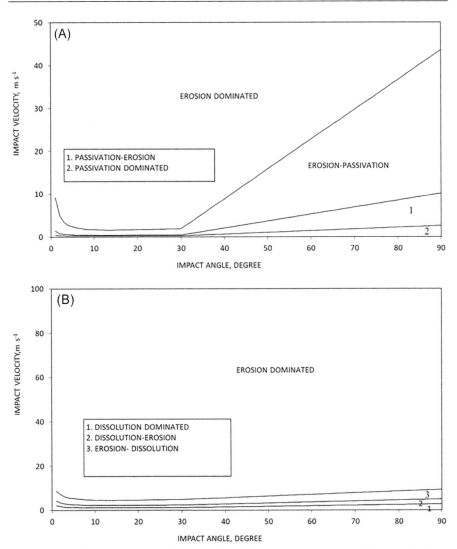

Fig. 9.9 Particle impact velocity-impact angle regime maps for pH 9 at −0.45 V: (A) Fe, (B) Ni,

Continued

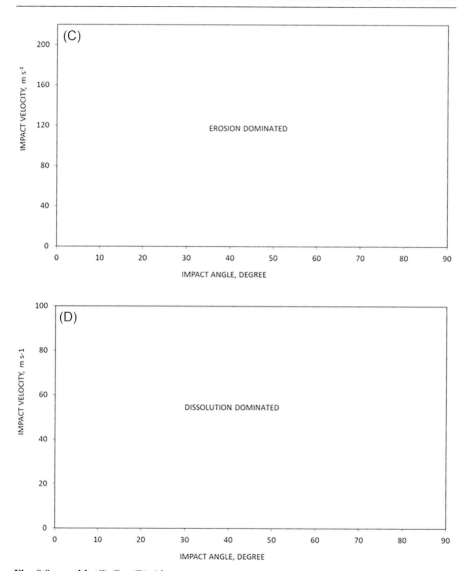

Fig. 9.9, cont'd (C) Cu, (D) Al.

Modeling erosion-corrosion in metals 217

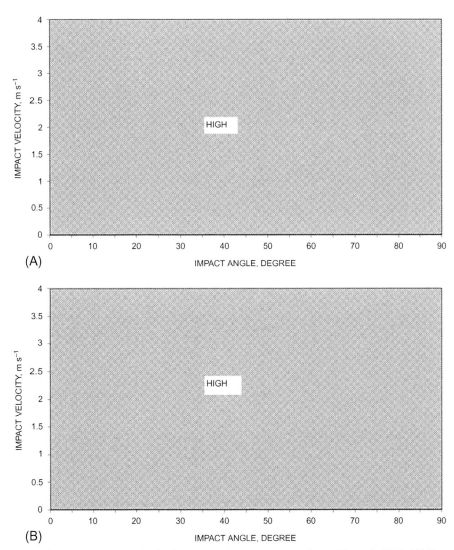

Fig. 9.10 Particle impact velocity-impact angle wastage maps for pH 5 at −0.45 V: (A) Fe, (B) Ni,

Continued

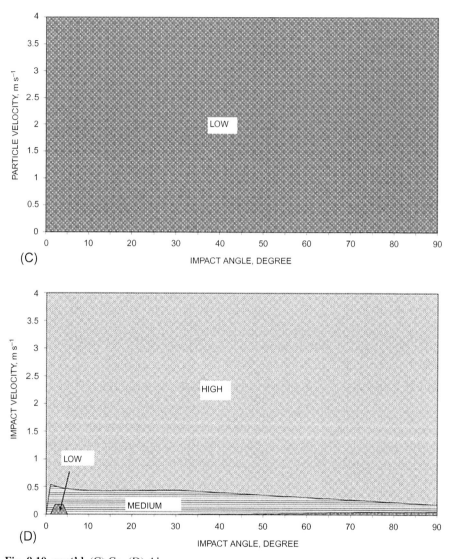

Fig. 9.10, cont'd (C) Cu, (D) Al.

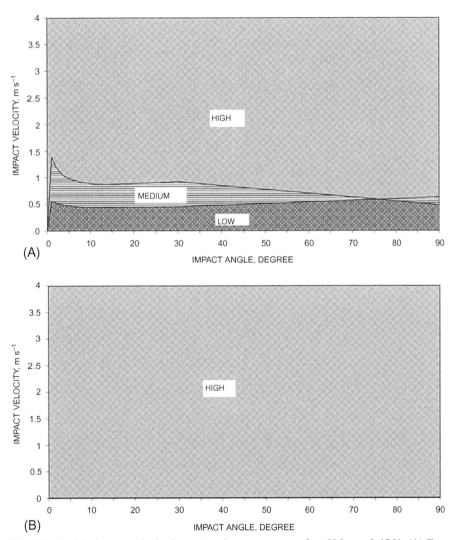

Fig. 9.11 Particle impact velocity-impact angle wastage maps for pH 9 at −0.45 V: (A) Fe, (B) Ni,

Continued

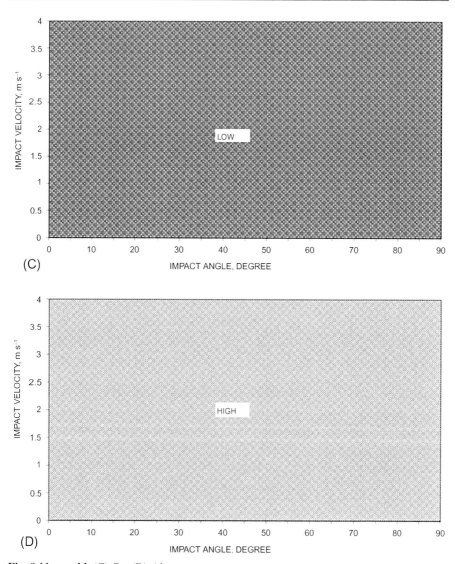

Fig. 9.11, cont'd (C) Cu, (D) Al.

The velocity-impact angle regime maps for pH 5 and 9 (Figs. 9.8 and 9.9) show significant differences in the erosion-corrosion response of the individual metal as a function of impact angle. At pH 5, both Fe and Ni demonstrate dissolution-assisted erosion-corrosion response, whereas, at pH 9, Fe shows passivation-assisted erosion response, but Ni remains unaffected by the changes of pH (from pH 5 to pH 9). Al shows passivation assisted erosion-corrosion response at pH 5, but its response

Modeling erosion-corrosion in metals 221

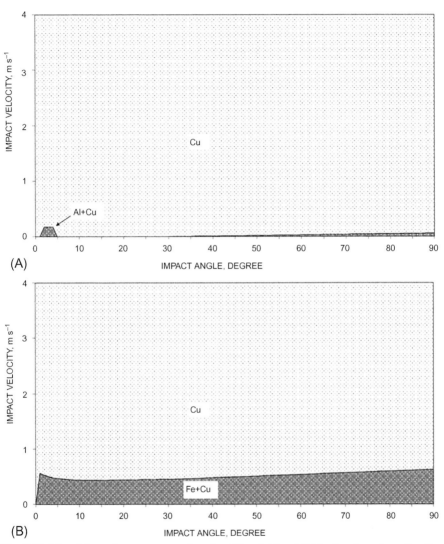

Fig. 9.12 Materials performance maps based on Figs. 9.10 and 9.11, showing where "low" wastage is observed for the various pure metals: (A) pH 5 and (B) pH 9.

changes to dissolution-dominated response at pH 9. Cu shows only erosion-dominated behavior at both the pHs and this is because of the high immunity of the metal in the condition modeled. At pH 5, the transition velocity to erosion-dominated response of Ni is marginally low whereas that of Fe is significantly high. Al, although relatively the most reactive metal, shows a transition between erosion and passivation-dominated behavior at lower pHs, this response is not evident in higher pHs.

Velocity-impact angle wastage maps for various metals at pH 5 and 9 are presented in Figs. 9.10 and 9.11. In Figs. 9.10 and 9.11 different shadings are used to highlight the "velocity vs. impact angle zones" of the low, medium, and high erosion-corrosion rates (KEC) as defined at the end of Section 9.5.2. Fe shows high wastage at pH 5, but the wastage rates seemed reduced at pH 9. The wastage regimes of Ni and Cu are unaffected by the changes of pH attempted in this study—Cu shows a low wastage regime whereas Ni shows a high wastage regime. Al shows a difference in wastage regimes at pH 5 and 9, with marginally lower wastage rates being observed at pH 5.

Velocity-impact angle materials performance maps for pH 5 and 9 are presented in Fig. 9.12A and B and these maps are generated by superimposing the low wastage regimes of the wastage maps of the various metals, Figs. 9.10 and 9.11. In the condition modeled, Cu shows high resistance to erosion-corrosion at both pHs. However, at pH 5, Al appeared to show a marginal resistance at lower impact angles, whereas, Fe, shows some resistance at pH 9.

9.7 Discussion

The erosion-corrosion response, as depicted in the various regime and wastage maps generated in this study, is significantly different for the individual metals studied. The influence of corrosion in the overall degradation is in accordance with the corrosion response as evident in the Pourbaix diagram of the various metal, Fig. 9.2 [4, 20].

In the velocity-potential regime maps, Figs. 9.3 and 9.4, Ni shows only dissolution affected region, whereas Fe shows both dissolution and passivation affected region in the potential range modeled. Albeit both Fe and Ni are active-passive transition metals, Ni passivated at significantly higher potentials than Fe and the passivation potential of Ni falls outside the potential range modeled. Therefore, this difference in the regime map of Fe, and Ni is not surprising. The maps also reveal that the pure dissolution (no erosion) regimes for Fe and Ni extend to a higher velocity at normal impact compared with that at oblique impact—this is partly because the erosion rate at oblique impact is significantly higher than that at normal impact (Fig. 9.1) and partly because for the consideration of the coefficient of restitution into the erosion model, Eq. (9.13), which predicts erosion rate lower compared to the model without accounting for elastic rebound effects [3, 4].

The regimes maps of Cu, Figs. 9.3 and 9.4, show the largest immune region where erosion is the only mode of degradation. At higher potentials, Cu, however, shows dissolution affected erosion behavior, which is in accordance with the corrosion response observed in the Pourbaix diagram of the metal. Figs. 9.5 and 9.6 and the materials performance maps, Figs. 9.7 indicate that Cu is the favored metal compared to the other various metals studied. It should be noted that the material performance map presented is only valid for the conditions modeled in this study and is likely to change if a different window of conditions is used.

The regime maps of Al, Figs. 9.3 and 9.4, shows passivation affected erosion-corrosion behavior, which is in accordance with the corrosion response observed in the Pourbaix diagram of the metal. Al, similar to Fe, shows comparatively higher transition velocity to erosion-dominated behavior at normal impacts than at oblique impacts. The wastage maps of Al demonstrate a relatively higher rate of the metal compared to the various metals studied and as a result, the material is less favored in the modeling conditions, Fig. 9.7.

The velocity-impact angle regime maps for pH 5 and 9, Figs. 9.8 and 9.9, show that the impact velocity required for the transition to the erosion-dominated regime is considerably lower at oblique impact angles than that at the normal impacts. This is expected as the erosion rate at oblique impact is higher than that at normal impact. The maps also show that at passivation affected conditions the transition velocity to erosion-dominated behavior changes at a faster rate with an increase in impact angle compared to the behavior observed in dissolution affected conditions. This may be since the corrosion rate in the passive conditions is more dependent on the size of the crater (which varies with impact angle) than in the dissolution affected conditions.

The velocity-impact angle wastage maps at pH 5 and 9, Figs. 9.10 and 9.11, show some interesting difference in the erosion-corrosion resistance behavior of Fe and Al—as the pH is increased from 5 to 9, the overall wastage rate of Fe reduces, but that of Al increases. This is because at lower pHs Fe degrades under dissolution affected process (consequently higher wastage rate), whereas Al degrades under passivation affected process (consequently lower wastage rate). This situation is reversed at higher pHs. Hence pH has a significant effect on these wastage maps of the various metals studied under the condition modeled.

The velocity-impact angle materials performance maps, for pH 5 and 9, Fig. 9.12, show that Cu has the highest resistance to erosion-corrosion degradation as compared with the various metals, therefore, Cu is the most preferred metal in the conditions modeled. Together with Cu, Al is favored at pH 5 for a specific condition involving a low potential and low impact angle and Fe is favored at pH 9 up to impact velocities less than 0.3 m s^{-1}.

Future work will to consider elastic rebound effects in a wider range of erosion-corrosion environments such as over a larger pH parameter space.

9.8 Conclusions

The elastic rebound velocity of erodent particles is shown to be significant for slurry erosion conditions, especially for low impact velocities (e.g., pipe flow of slurry). The effect of elastic rebound velocity and impact angle are modeled in erosion-corrosion environments and the influence of these variables on erosion-corrosion maps for various metals are explored. The rebound velocity has a role in extending the transition boundaries on the erosion-corrosion maps. The impact angle and pH are found to have major impact on the erosion-corrosion boundaries in the conditions modeled. The materials performance maps show the impact angle has a significant role on the optimum performance of the metals in the conditions modeled.

9.9 Notations

A	area of the projection of the crater form after particle impact (m^2)
C_p	specific heat of the target materials (J kg^{-1} K^{-1})
d	particle Indentation depth on target metal (m)
const.	constant for Eq. (9.35) (Dimensionless)
D_f	density of passive oxide film (kg m^{-3})
D_p	density of erodent particle (kg m^{-3})
D_t	density of target metal (kg m^{-3})
e	coefficient of restitution (Dimensionless)
E	dimensionless erosion (Dimensionless)
E_b	elastic modulus of erodent particle (Pa)
E_e	elastic modulus of collision (Pa)
E_{reb}	dimensionless erosion considering particle rebound effect (Dimensionless)
E_t	elastic modulus of target material (Pa)
f_i	fraction of particle concentration responsible for erosion event (Dimensionless)
F	Faraday's constant (C mol^{-1})
h	thickness of passive film form on the metal surface (m)
H_d	dynamic hardness of target material (Pa)
i_a	anodic current density (A m^{-2})
k	constant Finnie's model (Dimensionless)
K_C	total corrosion rate in the presence of erosion (kg m^{-2} s^{-1})
ΔK_C	additive effect of erosion affected corrosion (kg m^{-2} s^{-1})
K_{CO}	pure corrosion rate in the absence of erosion (kg m^{-2} s^{-1})
K_E	total erosion rate in the presence of corrosion (kg m^{-2} s^{-1})
ΔK_E	synergistic effect of corrosion affected erosion (kg m^{-2} s^{-1})
K_{EC}	total erosion-corrosion rate (kg m^{-2} s^{-1})
K_{EO}	pure erosion rate in the absence of corrosion (kg m^{-2} s^{-1})
K_{Ereb}	erosion rate considering particle rebound effect (kg m^{-2} s^{-1})
n	number of electrons donated by the dissolution reaction (Dimensionless)
r	radius of erodent particle (M)
RAM	relative atomic mass of target material (kg mole^{-1})
sol	percent solid by mass, expressed as a percent (Dimensionless)
T_m	melting point of the target material (K)
U_d	deformation velocity of target metal (m s^{-1})
U_i	particle impact velocity (m s^{-1})
U_r	particle rebound velocity (m s^{-1})
V_d	volume of metal deformed by a single indentation (m^3)
V_i	volume of indentation (m^3)
V_r	volume of relaxed crater (m^3)
V_{ur}	volume of unrelaxed crater (m^3)
w	diameter of crater from after a erodent particle impact (m)
α	incidence angle of impact of erodent particle on target surface (Degree)
υ_b	Poison's ratio of erodent particle (Dimensionless)
υ_t	Poison's ratio of target metal (Dimensionless)
ν_i	Stoichiometric coefficient for each reaction component (Dimensionless)
η	Over potentials (V)

References

[1] B.D. Jana, M.M. Stack, Wear 259 (2005) 243–255.
[2] M.M. Stack, N. Corlett, S. Zhou, Wear 214 (1998) 175–185.
[3] M.M. Stack, B.D. Jana, S.M. Abdelrahman, in: D. Landolt, S. Mischler (Eds.), Tribo-Corrosion of Passive Metals and Coatings, Woodhead Publishing, Cambridge, 2011, pp. 153–184.
[4] M.M. Stack, B.D. Jana, Wear 256 (9–10) (2004) 986–1004.
[5] M.M. Stack, N. Pungwiwat, Wear 256 (5) (2004) 565–576.
[6] M.M. Stack, S.M. Abdelrahman, B.D. Jana, Wear 268 (3–4) (2010) 533–542.
[7] M.M. Stack, S. Zhou, R.C. Newman, Mater. Sci. Technol. 12 (1996) 261–268.
[8] I. Finnie, 3rd US Nat. Cong. Appl. Mech, ASME, New York, 1958, p. 527.
[9] I. Finnie, Wear 3 (1960) 87–103.
[10] G. Sundararajan, P. Shewmon, Wear 84 (1983) 237–258.
[11] I.M. Hutchings, Proceedings of the 5th International Conference on Erosion by Solid and Liquid Impact, Cavendish Laboratory, University of Cambridge, 1979. paper 36.
[12] I.M. Hutchings, Wear 70 (1981) 269–281.
[13] M. Hutchings, Tribology: Friction and Wear of Engineering Materials, Edward Arnold, 1992.
[14] Y. Tirupataiah, B. Venkataraman, G. Sundararajan, Mater. Sci. Eng. A124 (1990) 133.
[15] D. Tabor, Hardness of Metals, Clarendon Press, Oxford, 1951.
[16] A.R. Singer, R.W. Evans, Metals Technol. 7 (1) (1980) 142–145.
[17] K.L. Johnson, Contact Mechanics, University Press, 1985 (Chaters 4 and 11).
[18] S.S. Brenner, H.A. Wriedt, R.A. Oriani, Wear 68 (1981) 169–190.
[19] I. Finnie, Wear 19 (1972) 81–90.
[20] M. Pourbaix, Atlas of Electrochemical Equilibria in Aqueous Solutions, National Association of Corrosion Engineers, Houston, TX, 1974.

A monitoring management strategy in tribocorrosion: Application to erosion-corrosion in oil and gas exploration-production

10

Antoine Surbled
Surbled Antoine Consultants, Paris, France

10.1 Introduction

Degradation of equipment in the energy industries is a serious concern actively addressed by corrosion management processes [1]. The erosion-corrosion mechanisms constitute a significant part of the potential degradation processes encountered in these industries and particularly in oil and gas exploration and production. One of the problems of exploration and production activities is the need to extract oil, gas, and water from reservoirs at ever-greater depths, where it becomes very difficult to intervene in the event of incidents despite advances in robotization.

The oil and gas produced from wells is made up of a multiphasic mixture which essentially consists of:

- liquid hydrocarbons (condensates, oils, bituminous), more or less loaded with corrosive compounds (dissolved H_2S, mercaptans, organic acids),
- solid hydrocarbon such as paraffins and waxes,
- hydrates,
- hydrocarbon gases (natural gas),
- other gases—hydrogen sulfide, carbon dioxide, nitrogen,
- water containing diluted salts, CO_2 and H_2S, and
- sand, particles of proppant, "black Powder"

The degradation mechanisms encountered are essentially due to the effect of corrosion and abrasion [2, 3] namely:

- corrosion (CO_2, H_2S, chlorides, MIC, and under deposits corrosion),
- stress corrosion cracking (H_2S),
- particles erosion (sand, proppants, and marginally other solids),
- erosion under the action of the shock of liquid droplets [4], and
- erosion-corrosion under combined action of corrosives (mostly CO_2) and erosive particles or droplets [5, 6].

But also:

- galvanic corrosion,
- microbiologically induced corrosion

It is generally accepted that particles (sand and proppants) are the most common source of erosion in hydrocarbon systems.

10.2 Erosion-corrosion in oil and gas production

An overview of the equipment used in the oil and gas production that is potentially affected by erosion phenomena, most of time associated with corrosion, is given in Fig. 10.1.

10.2.1 Erosion-corrosion concerns

Well, transfer, surface, separation equipment, gas flare, and compression systems are potentially subject to erosion-corrosion, until erosive and corrosive material are removed from the streams. Sea-bed facilities and flexibles are potentially subject to erosion.

Typically, erosion is a potential concern for all the parts located upstream of the last separator and in this separator, thus possibly in the seawater system, up to the filtration stage. Corrosion is a potential concern for all parts where water is present at least up to the driers.

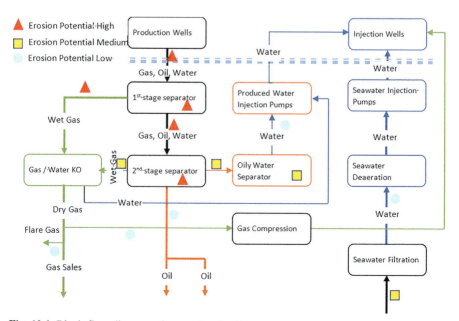

Fig. 10.1 Block flow diagram of separation facilities.

10.2.2 Erosive agents

The abrasive agents potentially present in the liquid and gas streams are of different kinds.

Produced sand is defined in terms of particle size between 62 and 2000 µm [7]. Particle erosion is governed by the fraction of quartz in the sand. The Mohs hardness of quartz is about seven, which makes it abrasive for most metallic alloys and composites. Particles smaller than 62.5 µm are called fines, there are generally less erosive than sand due to their small size and generally lower quartz content. Particles larger than 2000 µm are classified in the gravel category.

Barrite and calcite: Barite and calcite are used to densify drilling fluids and prevent gas leaks. Their Mohs hardness is around three, they are much less erosive than quartz particles. The size of the barite and calcite particles is around 20 µm. Under normal operating conditions, their abrasive potential is low, it can become high under certain conditions of a combination of high bulk velocity or high-local velocity and a large quantity of materials.

Proppants: Proppants (natural or synthetic) are solids, they are injected in suspension in the fracturing fluid into cracks and microcracks of the rock caused during hydraulic fracturing. They promote the formation of a permeable and resistant layer to keep the microcracks open after having penetrated them. The fluids percolate until then extracted through this layer. Proppants can be as erosive as sand, primarily during well clean up.

Drill cuttings are fragments of solid material from a borehole brought to the surface in drilling mud. Drill cuttings are formed when the rock is broken by the drill bit, they are transported to the surface by the drilling fluid that rises from the drill bit, drill cuttings can be erosive. When mixed with high viscosity mud, they are less so.

Corrosion or formation products such as "black powder" or scales such as sulfur particles. Black powder is a generic term for chemical compounds found in pipelines, and its composition is highly dependent on gas composition. According to Sherik et al. [8], in sour gas pipelines black power is mostly constituted of FeS, FeS_2, minor amounts of Fe_3O_4, FeO(OH), and $FeCO_3$, (siderite). In sales gas pipelines Fe_3O_4, α-FeO(OH), γ-FeO(OH), and small amounts of $FeCO_3$ have been found. Yamada et al. [9], found in the Japanese sales gas network, predominantly Fe_3O_4, small amounts of hematite (α-Fe_2O_3), goethite [α-FeO(OH)], lepidocrocite [γ-FeO(OH)], and $FeCO_3$. Generally, erosion by black power is not a concern, nevertheless, some cases are documented [10].

Shales are fissile rocks formed by the consolidation of clay, mud, or silt, they have a stratified structure and are composed of minimally transformed minerals. Shales are rarely erosive (Mohs hardness ≤ 3).

Nonerosive agents: Nonerosive agents are solids that do not have a hardness or size sufficient to cause ductile deformation of the surfaces or fatigue. Particles with a hardness of less than 3 Mohs are normally nonerosive for steel. Typically, chlorite, illite, illite-smectite, mixed layer, kaolinite, clays, silts (hardness 2.0–2.5 Mohs) are not erosive in fluids with a bulk velocity of less than 100 m s^{-1}.

10.2.3 Corrosive materials, CO_2 corrosion

CO_2, H_2S, water, hydrolysable mineral salts and bacteria promote the development of many corrosion phenomena in the production facilities and primary treatment section of mineral hydrocarbons.

The main forms of corrosion encountered are CO_2 corrosion, alone or in presence of H_2S. Microbiologically induced corrosion forms under corrosion products. Forms of stress corrosion may also develop. The corrosion mechanisms induced by these compounds are widely described in specialized literature and technical guidelines [11]. The CO_2 corrosion mechanism takes several forms such as mesacorrosion, pitting, mostly located at the bottom of equipment and piping, and a specific form in wet gas called the top of line corrosion, located on top of piping. CO_2 corrosion may have a little synergistic effect with abrasion [12, 13]. CO_2 corrosion occurs on carbon steels. Stainless steels and nickel alloys are immune.

Since the first models proposed in the 1970s by C. De Waard and M.D. Milliams [14], the mechanisms of this form of corrosion became better understood and current models take into account many parameters, including the presence of solid particles in the flux, a selection of some representative studies can be found in references [15–19].

Carbon dioxide (CO_2) in gaseous form is not corrosive. When it dissolves in water it hydrates (about 0.2% under normal pressure and temperature conditions). Two classic mechanisms are proposed for the dissolution of iron in acid solutions, namely the "catalytic mechanism" and the "consecutive mechanism." They are associated with two distinct electrochemical behaviors observed in the active dissolution range [15]. The catalytic mechanism was proposed by Heusler et al. [20]. It is based on the Tafel experimental slope of 30 mV and the second order dependence on the concentration of hydroxide ions. The consecutive mechanism proposed by Bockris et al. [21] was formulated to explain the Tafel slope of 40 mV and a first order dependence on the concentration of OH^- ions.

Keddam et al. [22] proposed a mechanism comprising seven steps resulting from three parallel dissolution routes incorporating elementary steps for the consecutive and catalytic mechanisms. A significant effect of carbonate species on the rate of dissolution of iron is mentioned in several studies [23, 24]. Kahyarian et al. [15, 25] summarize the actual knowledge regarding CO_2 corrosion. They suggest that the increase in the corrosion rate of steel in acid solutions in the presence of CO_2 is due to the increase in the dissolution rate of iron in the presence of CO_2, added to the increase in limiting cathodic current resulting from the buffering capacity of dissolved CO_2 and H_2CO_3.

10.2.4 System and subsystem components sensitive to abrasion and erosion-corrosion

Elements of production systems sensitive to erosion and erosion corrosion are essentially [2, 3]:

- downhole sand control devices (e.g., sand screen),
- downhole flow restriction devices (such as subsurface safety valves, emergency shutdown valves, inflow control devices, sliding sleeves, etc.),

- flexible pipes with interlock carcass,
- christmas tree, swivels, and flow line,
- choke valves and piping of downstream of choke valves,
- piping bends [26, 27],
- multiphase riser pipelines [28–30],
- flow restriction (Coriolis meters, some other types of flow meter) [31–33],
- sand separation cyclones,
- gas export line with solids carryover [34, 35],
- flare system with solids carryover, and
- water reinjection system with solid carryover.
- Flow disturbance location (e.g., weld bead, Tee, etc.)

Due to the small impact angle of the particles, erosion is generally low in the straight and nonrough parts of the pipes.

The erosion resistance tests must meet the conditions of section 7.7.7 of API 17B [36]. Based on current techniques, the tolerable erosion of the interlock casing must be limited to a maximum of 10%–30% of its thickness for the specified service life of the hose. Given a characteristic carcass thickness of about 1 mm, the erosion tolerance will generally be 0.1–0.3 mm. This tolerable erosion is less than that of rigid steel pipes. The erosion potential must be assessed for the part of the hose supposed to have the worst combination of the radius of curvature versus service conditions.

As a first example, carcass erosion of flexible with the internal carcass occurs when particles contained in fluids collide with its inner wall. In gas lines, the velocity of solid particles is high so in the presence of sand, the conditions are favorable to erosion. Corrosion of the carcass at the interface of the end fitting does not damage the sealing barrier or the locking mechanism. Corrosion of steel layers causes their gradual destruction and reduces resistance and fatigue life.

Another example is production chokes. According to available failure statistics, such choke valves are the components of oil and gas production systems that are the most susceptible to erosion. This is mainly due to the high flow velocities created by the vacuum in the valve. In addition to the risk associated with erosion, the choke valves can become clogged due to the accumulation of sand of dimensions larger than the passage. Specific erosion models have been proposed for chokes valves [2, 37–39].

10.3 Erosion and corrosion management strategy

Recently the connection between erosion and corrosion was made and the synergy between these two aspects of degradation was considered [30, 40–43]. In practice, most often, operators consider the erosion management strategy (included in sand management) and the corrosion management separately. Sand (abrasive particle) can cause erosion problems and loss of production due to the need for regular cleaning of the separators and is an important risk factor for well integrity, while the presence of corrosive compounds (in particular CO_2 + water), causes corrosion of carbon steel. The synergy between these two aspects is important, which means that they should be brought together. The erosion and corrosion management strategy belong to a multidisciplinary team that is responsible for determining the causes of the production of sand and corrosion and for developing a strategy adapted to this objective. The team is most generally formed of specialists from different disciplines and led by a coordinator. The role of the team is to assess the situation of the installation in question vis-à-vis of the risks linked to the associated degradations.

The erosion-corrosion management strategy should be defined from the start of field development [front-end engineering design (FEED) or detail engineering] in order to ensure appropriate sizing and selection of all the equipment and instruments necessary for activities relating to erosion-corrosion management (see Fig. 10.2). The strategy should reflect both erosion-corrosion control and monitoring methods and provides appropriate safeguards to control the risk. Erosion-corrosion management strategy should be an integral part of integrity management.

The final objective is to maintain the integrity and availability of installations at an acceptable level for all the actors during a determined period or throughout the project life [44] (including the extension of life that could be requested).

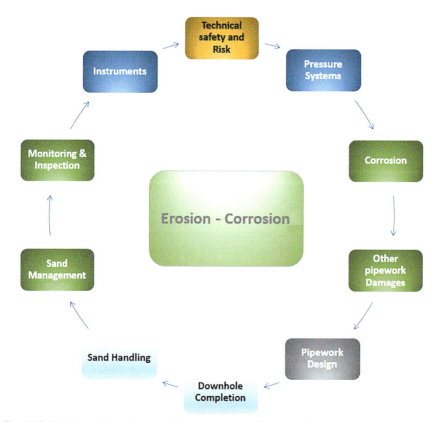

Fig. 10.2 Erosion and erosion-corrosion management framework.

The management of sand production, combined with the management of corrosion, should make it possible to optimize the production of the various wells and surface installations by avoiding production losses without compromising the safety and reliability of the installation. The whole process of sand management is described in technical and normative documents [2, 3, 45–49]. Therefore, the erosion-corrosion management details, corrosion and erosion models, control of erosion-corrosion, monitoring techniques, and methods, are solely considered hereafter. Fig. 10.3 recap the different stages of the sand and erosion-corrosion management processes.

The study of the reservoir (subsurface pool of hydrocarbons contained in porous rock formations) and its properties is the mandatory preliminary step and the decisions that will be made as to the methods of completion will depend on the erosion-corrosion management strategy. The erosion-corrosion aspect is interdependent on the project life and the technological choices adopted. During the preliminary phase of the project, the treatment of erosion-corrosion consists in exploiting the reservoir simulation results, selecting and exploiting the laboratory test results, choosing the models, anticipating the damage, and possibly, modifying basic technological choices or actions to mitigate the damage. This must allow the selection of an appropriate design of the components and subcomponents, and the selection of a proper corrosion inhibition system. Finally, a corrosion and erosion monitoring and inspection during project service life and corrective actions are required.

10.3.1 Tolerable erosion and corrosion

From a practical point of view, it is difficult to give no tolerance for erosion, it is customary to specify a minimum value. An acceptable rate of erosion and corrosion must consider the project life and the complexity, feasibility, and cost of repair or replacement, for example, it is very difficult or impossible to repair or to replace blow-out preventers (BOPs) located in the deep sea. On the other hand, the replacement of a hydrocyclone or a pump located on the floor is possible.

For steel pipes, tolerable erosion must be identified with reference to one of the following options:

- the minimum corrosion erosion tolerance must be at least equal to the accuracy of thickness measurement systems (in the order of 0.5 mm for an ultrasonic thickness measurement system),
- erosion corrosion tolerance is a function of the pipe class specification (PCS),
- the corrosion erosion tolerance must consider the corrosion component, for pipes lined with a corrosion resistant alloy, the tolerance can be a percentage of the lining thickness,
- tolerance for corrosion erosion identified on the basis of the minimum wall thickness required by the applicable specifications, and
- the corrosion erosion tolerance is defined according to the nominal pressure of the system (an analysis of the stresses in the pipes may be required for this option).

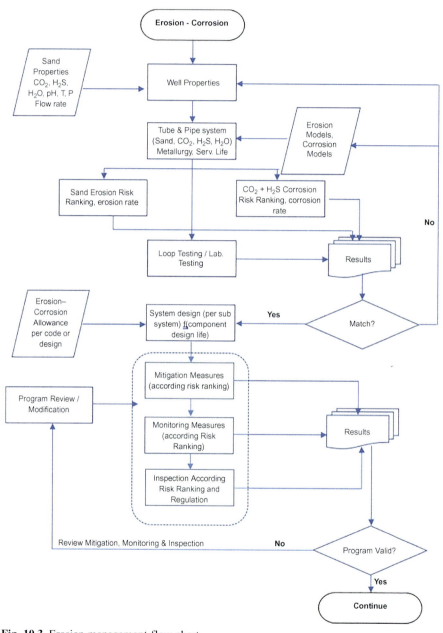

Fig. 10.3 Erosion management flow chart.

Decommissioning of the system to increase tolerable corrosion erosion should only be considered as a last option and should be subject to an in-depth assessment, also considering the future operations. For special equipment, the erosion-corrosion tolerance is generally provided by the manufacturer or the vendor based on tests and considering the potential effects on functionality.

10.3.2 Risk assessment

A risk assessment should be carried out to identify both the threats and the opportunities associated with sand erosion and corrosion. Risk assessment to set up and monitor a program providing sufficient, effective, and achievable guarantees for assessing, detecting, and managing sand erosion and corrosion.

10.3.2.1 Class of erosive service

A classification of the erosion potential of the pipes is defined in the document DNVGL RP O501 [2]. It refers to the average flow velocities calculated in the pipes, considering the correlation between the flow velocity and the intensity of the erosion. The erosion service class for piping is only an indicator of erosion susceptibility. The quantity and size of particles, the properties of fluids are not considered, which means that the class of service is not fully representative of real erosive potential (Table 10.1).

Table 10.1 Class of erosive service.

Erosion class	Pipework bulk flow velocity (m s^{-1})	Definition	Description
6	50–70	Extremely high erosion potential	System needs to be operated close to sand free. Safeguards to monitor erosion should be in place and closely monitored
5	30–50	Very high erosion potential	Tolerable sand production limited by risk of erosion
4	20–30	High erosion potential	Tolerable sand production will in most cases be dictated by erosion rather than sand handling capacity
3	10–20	Medium erosion potential	Tolerable sand production may be limited both by erosion and sand handling capacity
2	5–10	Low erosion potential	A large amount of sand is required to cause erosion. The acceptable sand load will in most cases be limited by the sand handling capacity in the process system.
1	0–5	Extremely low erosion potential	Effect of plain erosion, i.e., not considering any combined effects of FAC, cans normally be neglected for realistic sand loads

Based on DNVGL-RP-O501: 2015, Managing Sand Production and Erosion.

10.3.2.2 Risk assessment

The risk assessment and classification should be performed based on a breakdown of the system into subsystems and by specialists in each discipline concerned. The DNVGL RP O501 suggests a typical subdivision of systems into subsystems. For each of the subsystems, the risk associated with erosion-corrosion should be assessed on a qualitative basis regarding risk categories (for example those suggested in Table 10.2).

The risk assessment should address:

- critical system functions,
- consequence of each failure mode related to sand,
- acceptance criteria related to containment, function, availability, and performance,
- safeguards—risk reducing measures, and
- risk level with and without active safeguards.

The risk associated with erosion-corrosion should be made visible both with and without safeguards to emphasize the importance of following up both active and passive safeguards.

10.3.3 Basis of the sand production strategy

Sand production is associated with geological factors and, therefore, if a well in a field is known to produce large amounts of sand, other wells are suspects. The collection of sand in the separators is sometimes used as an indication of sand production. The parameters related to the sand management and in general of the particles contained

Table 10.2 Risk ranking.

Risk ranking	Definition	Action
Intolerable	Risk is in conflict with acceptance criteria or noncompliance with standards or regulations. System cannot be operated without modifications to system or operating procedures, or with additional safeguards	Review and correct
High	Risk could be acceptable in certain conditions. Systems need to be reviewed and detailed risk analysis shall be performed, generally modification of design or additional monitoring of safeguards shall necessary	Review and improve
Medium	Risk acceptable. Additional monitoring or safeguards shall be evaluated by operator according to ALARP principle	Improve monitoring and safeguarding
Low	Low risk with current operational procedures and safeguards	No action required

Based on DNVGL-RP-O501: 2015, Managing Sand Production and Erosion.

in the liquid or gaseous streams coming from hydrocarbon reservoirs are described in professional reference rules in particular API 14E [45]. This API 14E is widely used during the design of offshore oil and gas installations as well as in the DNVGL-RP-O501 [2] guidelines on sand erosion and erosion-corrosion management [3] and technical communications [50–53]. The recommendations issued by these documents are based on numerous scientific and technical publications. Fig. 10.4 summarizes the result of the application of a sand management strategy.

The sand management strategy, including erosion management strategy and corrosion management includes many possible measures.

10.3.3.1 Sand control

The control of sand production includes all the methods aimed at controlling its production in wells as well as the methods aiming to limit its effects, i.e., monitoring erosion and mitigation measures. Detail of sand control strategy is developed in specialized articles or books such as [54, 55]. Sand control includes the following aspects:

- prediction of sand production from geological and mechanical rock data,
- consequence assessment of sand production with and without using screens,
- avoiding sand production,
- coping with sand production,
- sand detection (monitoring),
- evaluation of the size distribution of the grains of sand,
- use of adapted screens, and
- chemical consolidation (sand consolidation, resin coated sand).

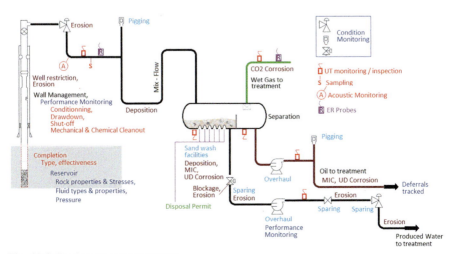

Fig. 10.4 Sand management strategy.

10.3.3.2 Selection criteria and methodology

The process involves the management of wells. Hereto, it is necessary to undertake a review of the history concerning similar wells if they exist then to analyze all the data relating to the production of sand and corrosion. Based on the existing data, models or the exploitation of test loops results, the characteristics relating to sand and corrosive materials can be defined. The harmfulness of erosive particles and corrosives in existing wells in service and those in development shall be assessed, the key decision elements are as follows:

- what are the foreseeable failures in the casing of wells and surface equipment? These potential failures are identified in terms of health, safety, and environment (HSE), public image, costs associated with repairs, replacements and production losses,
- size of sand particles, the selection of sand control is partially based on the dimensional distribution of particles in the reservoir, in general coarse grains are not brought to the surface, and
- corrosivity assessment of stream, the corrosivity of stream is depending of the nature and the amount of corrosive species especially CO_2 and H_2S.

The drilling conditions can affect the methods of controlling sand production and the impact on disturbances. The presence of large shale sections along the wellbore can create problems when drilling wells (difficulties in cleaning gravel). The characteristics of the reservoir formation (net to gross), permeability, number of zones, pressure, depletion levels, and temperatures will all have an impact on the selection and design of sand control. Production performance and reservoir management need to be evaluated based on the oil, gas, and water flows.

10.3.3.3 Sand production mitigation measures

Mitigation measures reduce and control the effects of sand production on the degradation of materials and optimize the operation and production of wells.

The measures taken can be of several types. A reduction in the production rate lowers both the production of sand and the flow velocity through the piping. The financial implications are obvious, and generally, this reduction of production is considered as ultimate when other options do not satisfy the objectives. The sand production and sand transport mitigation is the first step of sand management and strongly influence erosion and corrosion management. Screens may be installed at the bottom of the well, they may consist of gravel blocks that prevent the sand from entering the production system. Typically, sand screens prevent particles larger than 100 μm from entering the workflow, however, they increase resistance to flow and therefore affect the potential productivity.

Screens are used to prevent the sand produced from rising to the surface. They are of three types [55], namely: (i) wire-wrapped screens (WWS), (ii) prepacked screens (PPS), and (iii) mesh screens.

Since the 1990s, gravel packs are commonly used as filters, especially on horizontal drilling, the annular space between the gravel prevents the passage of sand, their success is not always certain.

Hydrocyclones and other types of sand collectors can also be used to reduce erosion of components downstream from the wellhead.

The installation of a series of inserts specifically calibrated in the flow makes it possible to reduce the effects of concentration of solid particles in the flow. These inserts can be installed during pipe cleaning operations.

10.3.3.4 Material selection and corrosion inhibitors

The erosion-corrosion simulation is always carried out initially by considering carbon steel as basic material. When carbon steel does not meet the criteria imposed by the project (degradation rate vs. service life), it is necessary to select materials more resistant to erosion or corrosion. It is possible to use corrosion-resistant alloys (CRAs) in order to limit corrosion. CRAs do not have a great effect on the erosion component, CRAs are typically austenitic or austeno-ferritic stainless steels, nickel alloys such as alloy 625, the selection of the appropriate alloys is carried out based on recommendations from technical documents having a normative value near the authorities and insurance and classification companies, NACE MR0175/ISO 15156 [11], NORSOK M-001 [56], and ISO 21457 [57]. Special heat treatments can also be used.

"Hard" alloys are used to limit the effects of erosion on the sensitive elements of the subcomponents (valve seats), they are essentially WC-CoCr cermet's, WCoCr alloys, they are also selected based on the same recommendations as to the CRAs [58–60].

Corrosion inhibition is currently used to mitigate the effects of corrosion on carbon steel materials. The different inhibition mechanisms are summarized in publications such as [61–65]. The film-forming corrosion inhibitors form a protective film on the metal surface due to physical (electrostatic) adsorption, chemisorption, or π-orbital adsorption, they can also be classified as anodic, cathodic, or mixed based on the type of reactions they inhibit. Imidazoline-based inhibitors are widely used to inhibit CO_2 corrosion. They generally consist of three parts: an imidazoline ring, a side chain, and a long hydrocarbon chain. Nitrogen atoms are the adsorption sites; the side chain blocks corrosive ions and molecules; the hydrocarbon chain, generally hydrophobic, can drive the molecule toward the interface and act as a barrier against water molecules and other corrosive species. Other inhibitors such as "Quat type inhibitors," acetylenic type inhibitors, polymeric inhibitors, phosphate inhibitors are also used.

Environmentally friendly inhibitors which are made from natural plant tissue extracts are being developed.

Monoethylene glycol is injected into pipelines to prevent the formation of hydrates and a subsequent blockage of the pipelines.

10.3.4 Experimental methods for investigating erosion-corrosion

10.3.4.1 Erosion-corrosion laboratory and loop tests

Erosion and corrosion simulation can be performed in the laboratory or on test loops. Basic electrochemical methods such as electrochemical potential monitoring, polarization measurement, electrochemical impedance spectroscopy, electrochemical noise [66, 67] can be used to determine the effect of corrosion. The main methods used for erosion detection are rotating disc/cylinder systems, slurry jet impingement rig systems, test flow loop wire beam electrode coupled with 3-D profilometry, and the lab electrical field mapping technology.

Among the rotating electrodes system, the most widely used are:

- rotating disc electrodes (RDE) that are simple tools where the working electrode current depends on the square root of the angular velocity,
- rotating ring disc electrodes (RRDE) used for the quantitative and qualitative identification of the species dissolving on the disc-shaped working electrode,
- rotating cylinder electrodes (RCE) which provide uniform shear stress on the cylindrical surface,
- rotary cage (RC) consisting of a sample holder for rotary coupons - weight loss data mainly used for the selection of inhibitors.

The test sets consist of a three-electrode system and a disc (RDE/RRDE) or rotation cylinder (RCE) [68] to simulate a hydrodynamic condition. In the RDE tests, the test sample is in rotation, it can also be placed at the edge of the rotating disc (RDDE). In the first case, the linear velocity on the surface of the sample can be controlled by the speed of rotation. In the second case, the flow rate of the slurry suspension on the surface of the sample cannot be directly related to the linear velocity of the disc due to the side effect. As the angle between the sample and the direction of flow cannot be adjusted, loss of metal due to erosion in the first case comes from the tangential direction at the start of the test. However, the angle between the sample and the flow direction can be adjusted from 0 degrees to 90 degrees in the second case. Consequently, the influences of the angle of impact of the sand can be roughly studied using the second set.

The hydrodynamic parameters of the rotary cylinder system are known. The Reynolds number (Re) and the wall shear stress (τ_{RCE}) can be calculated on the basis of theoretical and empirical equations.

$$Re = \frac{\omega r_{RCE} d_{RCE}}{\upsilon} \tag{10.1}$$

$$\tau_{RCE} = 0.079 Re^{0.30} \rho_m \omega^2 r_{RCE}^2 \tag{10.2}$$

where ω is the rotation speed in rad s^{-1}, r_{RCE} is the radius of the RCE, d_{RCE} is the diameter of the rotating cylinder electrode, υ is the kinematic viscosity of the electrolyte, and ρ_m is the density of fluid.

The rotating disc/cylinder system simulates an erosion-corrosion environment. There are however some system limitations such as:

- control of the concentration of particles when introducing the slurry into the test cell. Although the overall percentage of particles is known, the actual concentration of particles at different levels of a liquid varies due to the nonhomogeneous distribution in the cell. The samples are fixed at a certain height, the actual concentration of particles at the corresponding liquid level varies with the variation of the speed of rotation, and

- the RDE or RCE systems do not allow the study of erosion-corrosion under a single particle impact. The working environment is a pure electrolyte or a suspension in the test cell. As the system is completely sealed, the particles in the test cell cannot be replaced when the sharp edges of the particles are worn, changing the shape of the particles changes the results.

Slurry jet impingement rig systems can be used for the evaluation of pure erosion and erosion-corrosion. The system consists of a measurement and control unit (electrodes, pH measurement) and a mud loop driven by a pump. The suspension spray impacts platform provides control of the impact velocity and impacts angle of the particles. The particles can be injected into the flow loop using a hopper which can adjust the intensity of the impact of the particles. Erosion-corrosion can be assessed under the impact of a single particle.

The sample is always placed a few millimeters below the reaction platform to maintain a stable impact velocity. The angle of impact can be adjusted by rotating the sample holder. The distribution of the flow below the jet platform is not homogeneous, which implies that the wall shear stress, the angle of impact, the velocity of impact, and the trajectories of particles at local surfaces are different. The angles of impact of the particles are influenced by the flow of fluid. The center of the sample undergoes a normal impact which gradually changes into an oblique impact with the increasing distance from the center of the sample. The main types of the test apparatus are described in the paper [69].

Test flow loops are used to simulate the erosion-corrosion behavior of complex pipe sections composed of elbows, welds, and diameter variations. The test loops are instrumented to allow the monitoring and recording of erosion-corrosion data, the results are closer to reality than those obtained by other laboratory methods. It can be used to study the inhomogeneous erosion-corrosion behavior around the pipe circle using a designed sensor system. A test loop is made up of a water tank, a centrifugal pump designed for service in the presence of abrasive particles, flow, and pressure control instruments. According to the hydrodynamic parameters measured, the distributions of the flow velocity, the wall shear stress, and the concentration of sand at a certain section of pipe, can be predicted from computational fluid dynamics (CFD) simulations.

The wire beam electrode (WBE) in combination with a three-dimensional (3D) surface profilometry measurement was proposed by Xu et al. [70, 71]. It allows to measure the localized corrosion rate under dynamic flow conditions and to simulate the progression of flow accelerated corrosion (FAC) and erosion-corrosion on a steel support.

The WBE technique makes it possible to restore the electrochemical integrity and the change of the local chemical environment on a monoblock sample. At the same time, the total loss of metal from a single electrode can be measured from the loss of volume using 3D profilometry.

The lab electrical field mapping technology unit consists of two components, namely an electric field mapping acquisition (EFM) unit and a constant current source.

The base system works with sensor arrays up to 128 pins by adding additional analog multiplexer modules. The temperature is compensated. The various control parameters are entered by the user via a software interface.

10.3.4.2 Erosion and corrosion models

Commonly used particles erosion models consider four stages in the erosion process:
1. modeling of the flow of the carrier fluid through the element considered,
2. prediction of the flow to deduce the drag forces generated by the fluid on the particles and from there, prediction of the trajectories,
3. evaluation of the damage caused to the wall by the impact of individual particles using a chosen model, and
4. consideration of the particle distribution over a large number of possible trajectories, summation of the results by impact position.

Modeling erosion requires the selection of a suitable and accurate model. In the oil and gas industry, impact damage models have a similar basis. However, a number of methods are used to calculate the trajectories of particles, Arabnejad et al. [72] and Abdulla [73] have evaluated solid particle erosion and models.

According Mohyaldin et al. [33], erosion models could be classified into three categories, namely empirical models, mechanistic models, and computational fluid dynamics models.

The **simplest erosion model** used when sizing production systems is given in API 14E [45], it defines an acceptable average flow velocity (no particles or bubbles tracking) of the pipeline as:

$$V_e = \frac{C}{\sqrt{\rho_m}} \tag{10.3}$$

in which:

where V_e is the erosional velocity in ft s^{-1}, ρ_m is the density of the gas/liquid mixture at flowing pressure and temperature (lbs ft^{-3}), C is an empirical constant (which takes the value of 100 for a corrosive system and 150–200 for an inhibited system), the conversion factor between USCS and SI units is 1.21, other values have been proposed by many operators, the advantages and disadvantages of this method are discussed in several publications [74–77]. This method is still used for initial material selection.

The model proposed by Huser and Kvernvold [78] is the basis of DNVGL RP O501 model and uses the following form impact damage equation:

$$E_m = m_p K U_p^n F(\alpha) \tag{10.4}$$

In which E_m is the actual material loss rate in kg s^{-1}, m_p is the mass rate of particles in kg s^{-1}, $F(\alpha)$ characterizes the ductility of the target material, U_p is the particle impact velocity in m s^{-1}, K and n are the constants defined for materials in steel, titanium, and GRP. The values of K, n, and $F(\alpha)$ were obtained from sandblasting tests.

Salama and Venkatesh [79], use methods similar to those of Huser and Kvernvold, although they simplify their model by making the conservative assumption that all impacts of sand occur at around 30 degrees and therefore they set $F(\alpha)$ to 1. This approximation is reasonable for gas flows, but does not take into account the effects of particle drag in liquid flows. Hence, the following equation:

$$ER = S_k \frac{WU_m^2}{PD_i^2} \tag{10.5}$$

ER is the erosion rate (thousandths of an inch per year), W is the sand production rate (lb day^{-1}), U is the fluid flow velocity (ft s), D_i is the internal diameter of the pipe (in.), S_k is a dimensionless geometry-dependent constant which equal 0.038 for short radius elbows and 0.019 for ells and tees, P is the hardness parameter (bar), for steel pipes $P = 1.05 \times 10^4$ bar, In SI units, assuming the sand density of 2650 kg m^{-3}, can be written:

$$ER = 37.585 \frac{WU_m^2}{PD_i^2} \tag{10.6}$$

Svedeman and Arnold [80] also suggested using the equation of Huser and Kvernvold. Based on the correlations developed by Bourgoyne [81], with S_k values for gas systems: $S_k = 0.017$ for long radius bends and ells $S_k = 0.0006$ for plugged tees.

In this model, the sand particle size, shape, and hardness are neglected, the model is not applicable to two-phase flow. This method has proven to be constantly conservative, the expected erosion rate being on average 44% higher than the measured value.

In 2000, Salama [82], incorporated the effect of two-phase mixture density and particle size and proposed:

$$ER = \frac{11.574 WU_m^2 d}{S_m \quad \rho_m D_i^2} \tag{10.7}$$

where W is the sand production rate (kg s^{-1}), d is the particle diameter (μm), D_i is the pipe internal diameter (m), U_m and ρ_m are the mixture velocity (m s^{-1}) and density (kg m^{-3}), S_m is a geometry-dependent constant.

Calibration of the equation from experimental data from Weiner and Tolle [83] and Bourgoyne [81], leads to the constants $S_m = 2000$ for the 1.5D and 5D elbows (Bend radius 1.5 and 5 times the pipe outer diameter), $S_m = 12,000$ for ells, $S_m = 25,000$ for the plugged tees (liquid gas) and 500,000 for plugged tees (gas).

The **direct impact model** (DIM) is a mechanistic model developed by the Tulsa University Erosion/Corrosion Research Center (E/CRC) to predict penetration and direct impact rate for elbows and tees. Erosion is predicted using simplified particle trajectory equations, it is necessary to know the data relating to the pipe element (geometry and dimension), the flow (velocity, density, and viscosity), and the particles (density, size, and shape). To take into account the trajectory of the particles along with the flow, a concept of equivalent stagnation length has been introduced, this concept can be explained in the same way as equivalent length used to predict the local pressure drop in the fittings, in which different pipe element geometries have different equivalent stagnation lengths [72, 84].

The DIM model was used to develop a tool to predict the erosion rate [33].

DIM model can also be applied to multiphase flows (liquid/gas) by calculating the equivalent viscosity and density of the mixture [85].

Research conducted at Erosion/Corrosion Research Center of Tulsa University has led to a model widely used in the petroleum industry:

$$E = F_M F_S m_p U_p^n F(\alpha) \tag{10.8}$$

where E is the erosion rate (kg of material removed/kg of erosive particles), F_M is function of the hardness of the material according to McLaury and Shirazi [84, 86], its value would vary between 0.833 and 1.267 for various steels, F_S is the function of the angularity of the sand, its value varies between 1 (sharp grains) and 0.2 (rounded grains), m_p is the mass rate of particles in kg s^{-1}, U_p is the impact velocity of the particles, α is the impact angle, $F(\alpha)$ is a material-dependent function of the impact angle varying between 0 and 1, n is a material-dependent index. Haugen et al. [38] and Wallace [87] suggest that the variation in erosion resistance between different grades of steel is negligible compared to the spread of data points inherent in erosion tests. The models are often approximate, and the results obtained do not always correspond to the predictions.

Elbow erosion models for multiphase flows have also been developed by Birchenough et al. [88].

Computational fluid dynamics (CFD) simulates fluids in a dynamic state using numerical methods. The solution is obtained using models and techniques that are suitable for several applications. The CFD simulation of sand erosion is generally carried out in four stages:

1. constructing the geometric model and creation of a mesh by grid generation,
2. predicting the values relating to the velocity of the fluid along the direction of flow by resolution of a model of flow and one model of turbulence,
3. predicting the speed and angle of impact of particles using the particle equation of motion (Eulerian or Lagrangian), and
4. introducing particle velocity and impact angle data into a selected erosion prediction model.

10.3.5 On site erosion-corrosion monitoring

The rate of erosion-corrosion is likely to vary considerably over the time depending on the variation of surface properties, flow velocity, impact angle, temperature of the fluid, and the concentration of sand. Therefore, the erosion-corrosion rate should be monitored online to detect the behavior of erosion-corrosion.

Flow distributions and sand concentration are not uniform across pipelines. Different measurement results are obtained when erosion-corrosion probes are attached to various locations in a pipeline.

The methods of evaluation of erosion-corrosion rates used in laboratory such as electrochemical potential monitoring, EIS, and electrochemical noise are not sufficiently developed at industrial stage. Various types of sensors, data acquisition, and data processing systems are used when erosion-corrosion problems are suspected, they only provide an indication of the amount of sand produced and of erosion-corrosion rate, this is most often linked technological limitations and implementation difficulties on production sites (Fig. 10.5).

Monitoring devices may be located at the bottom of the well, on submarine pipes, on the production piping or line pipes, and upstream of the separators. The correct positioning of the probes is essential because the erosion of a component depends on the geometry of this component and the flow.

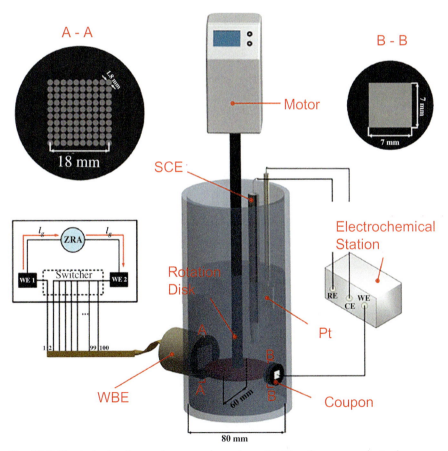

Fig. 10.5 The test setup for erosion-corrosion using a WBE and a coupon electrode. Reproduced with permission from Y. Xu, L. Liu, Q. Zhou, X. Wang, Y. Huang, Understanding the influences of pre-corrosion on the erosion-corrosion performance of pipeline steel, Wear 442–443 (2020) 203151, 0043–1648/© 2019 Elsevier B.V., Fig. 1.

The losses of thickness by erosion generally occur on the external radius of the elbows between 30 degrees and 90 degrees, however, the disturbances of the flow upstream of the elbow, can act to move the place of maximum erosion, which makes it difficult to detect. The loss of thickness by CO_2 corrosion occurs more generally on the areas where a water phase is present, generally on the bottom of the lines, or in case of top of line corrosion on the top of the pipeline.

Two types of techniques are used to monitor sand erosion and corrosion, namely:

- intrusive probes, inserted through the wall of the piping in the flow, and
- nonintrusive probes fixed to the piping wall,

Several detection methods are implemented, like:

- pressure variation in a sensor (erosion),
- acoustic measurements (noise detection) (erosion),
- detection of a loss of thickness based on the use of ultrasonic waves (erosion and corrosion), and
- detection of a loss of thickness based on the measurement of an electrical resistance (erosion and corrosion).

Corrosion and erosion monitoring methods are described in detail in technical guideline such as [89, 90] and technical publications [43].

In the case of intrusive probes, one end of the probe is attached to a high-pressure access fitting on the outside of the piping (Fig. 10.6). The other end located inside the pipeline and positioned judiciously in the flow.

A typically available technique is the sacrificial sand probes. The probe consists of a thin-walled, sealed tube made of a corrosion-resistant alloy and placed in the flow. When the sand comes into contact with the sealed element, it erodes it and when the wall is drilled, the system pressure is transferred to a pressure gauge or pressure sensor. The pressure device indicates that the thin-wall element has been broken. These probes do not give a direct indication of the erosion rate.

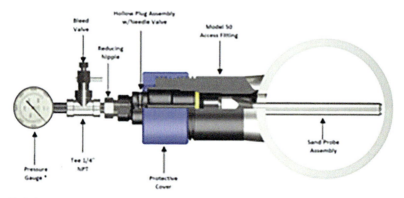

Fig. 10.6 Sand erosion detection system/pressure gauge assembly. Reproduced with permission from COSASCO.

This concept of the probe requires that the midpoint of the tube be placed approximately 60% in the internal diameter of the pipeline. This placement allows the maximum theoretical impact of the sand in the center of the line where the product flow velocity should be the highest, while also allowing the higher density suspended particles located in the lower part of the flow to also come into contact with the sacrificial tube.

Standard electrical resistance (ER) probes are used to monitor the total loss of metal caused by erosion-corrosion, they provide the cumulative loss of metal without having to remove samples from the service environment and can also be used in conductive and nonconductive environments. Their disadvantage lies in the fact that the sensitivity of the measurement strongly depends on the thickness of the sensitive element. Reducing the thickness of this element improves the sensitivity of the measurement but shortens the lifespan of the sensor.

The electrical resistance probes are intrusive, the detection part of the best-suited type to corrosion erosion detection has a rectangular shape. The probe is made up of a detection element exposed directly to the erosion-corrosion environment and a temperature compensation element that is sealed inside the probe to prevent any loss of metal (Figs. 10.7 and 10.8).

The resistance of each element can be measured using the four-wire method and the total corrosion depth x of the sensitive part can be expressed as follows:

$$x \propto t_k \left(1 - \frac{R^t_{ref}/R^t_{sen}}{R^0_{ref}/R^0_{sen}} \right) \qquad (10.9)$$

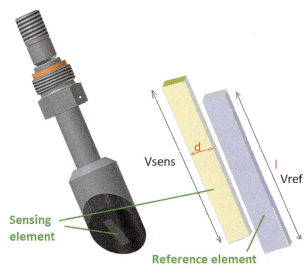

Fig. 10.7 Standard probes (adapted).
Reproduced with the permission from COSASCO.

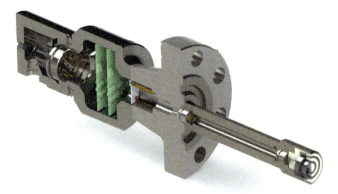

Fig. 10.8 Corrosion sensor (CEION Technology), cutaway view.
Reproduced with permission from Teledyne, Corrosion sensor CEION Technology.

where t_k is the thickness of the detection element, R^t_{ref}/R^t_{sen} is the actual resistance ratio between the reference element and the detection element, and R^0_{ref}/R^0_{sen} is the initial resistance ratio without corrosion or erosion.

The high sensitivity electrical resistance probes were developed in the late 1990s. In particular, the CEION Technology enabled a better resolution of the measurements. The main features of the instrumentation are the excitation method, the measurement of the response, strong noise filtering, a precision analog circuit, advanced digital signal processing, and temperature correction.

A high signal/noise ratio allows the detection on-site of a loss of 1–2 nm from a 0.5 mm thick sensor. Current systems compensate for small differences in the materials of the probes and thus obtain comparable results for the same level of corrosion or erosion.

Ring pair resistance electrical resistance sensors (RPERS) were developed by the end of the 1980s. The basic principle of these sensors is to divide the pipe into small sections, and the resistance of each section is monitored online. The localized erosion-corrosion rate of each segment can be related to the localized erosion-corrosion depth [55, 91]. These sensors are no more commercialized at the date, due to operational difficulties.

Nonintrusive probes are located on the outer wall of the piping. These probes are not affected by the process fluid, but close contact with the pipe wall is essential.

Acoustic-based nonintrusive sand monitoring sensors operate on the principle of detecting the sound created by the sand particles when they impact the pipe wall, this noise is then processed and used to identify and calculate sand production in real-time across all pipeline flows. The sensor is attached to the outside of the pipe, typically downstream of a 90-degree elbow. The particles of sand which are transported by the fluids strike the wall of the pipe immediately after the elbow due to inertia generate a high-frequency (100–500 kHz) acoustic pulse in the metal, the sensor through the microphone mounted on this end picks up the noise that propagates in the wall of the pipe and converts it into an electrical energy of impact digital signal (E_k) which is transmitted on the sensor's power cable to the calculation interface unit which calculates the production rate of sand based on the sensor signal and built-in algorithms. The noise generated is directly proportional to the velocity and the amount of sand in the fluid stream. As a result, the response is immediate and proportional to changes in sand production. The kinetic energy (E_k) is dependent on the impact velocity (v) and the mass of the grain (m):

$$E_k = \frac{1}{2} mv^2 \tag{10.10}$$

They operate at an ultrasonic frequency allowing them to filter background noise. The difficulty is to separate the noise generated by the impacts of the sand on the walls, that generated by the flow and the structural noises. Branded acoustic particle monitors can detect particles from a size of around 15 μm. The correct positioning of the probe is essential because the throttling noise can be much higher than the signal noise. Downstream of the constriction, the velocities, and therefore erosion are the highest. The sensitivity of sand probes in dry gas flows is around ± 0.1 kg day^{-1} and in oil flows around ± 0.1 kg/100 m^3. The calibration of the measurement system is most often carried out in situ by volumetric measurement of the sand flow in real-time for the flowing stream.

Piezoelectric ultrasonic thickness gauges are made from an ultrasonic array of sensors permanently placed to provide wall thickness readings over a section of pipe. Sensors can be arranged axial as a six-o clock position or arranged circumferential around the pipe section in order to measure/detect localized corrosion.

Guided waves monitoring uses transduction based on contactless EMAT technology to map the thickness of the wall of a pipe section [92]. The pipe wall operates as an ultrasonic waveguide. Each EMAT acts as a point source producing an undulatory cylindrical wavefront whose wavefront expands. As the wave front expands, it also wraps several times around the pipe, which leads to a propagation phenomenon with a helical path (Fig. 10.9).

For each source of the Tx, array waveforms are received by all the transducers of the network Rx (radial). Each waveform contains several pulses, each corresponding

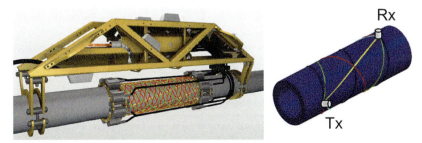

Fig. 10.9 Signal paths between multiple transducers on a subsea CEM System. Reproduced with permission from ClampOn, ClampOn Subsea CEM.

to a particular helical path. The different helical paths make it possible to build a thickness map with acceptable precision.

The **field signature method** (FSM) method makes it possible to assess the damage in steel structures from the results of measurement of the potential difference obtained by applying an electric current to a section of equipment [93–95]. In case of loss of section or crack on this steel section, the electrical resistance is locally increased at the level of the corroded or cracked area. According to Ohm's law, the potential difference increases in the proportion to the increase in electrical resistance. The detection matrix (SM) is made up of a series of studs welded to the area to be monitored. This fixed matrix offers a very high degree of repeatability in readings. A detection matrix interface (SMI) forms an interface between the SM and the data logger and includes the noise filtering circuits. The connection point of the instrument is located in a location that provides safe and practical access, generally, the cable has a length of 3–10 m (Fig. 10.10).

Many technical problems related to complex connectivity made it unreliable, submarine applications were quickly abandoned.

The **electrical field method** (EFM) method is close to the FSM method, but the pins are held by a cable tie, a two-part assembly, and a network of circumferentially assembled electrodes. The cylindrical head pressure maintenance screws are positioned on a fiberglass sleeve using special inserts. These screws are used to make electrical contact between the sensing electrodes.

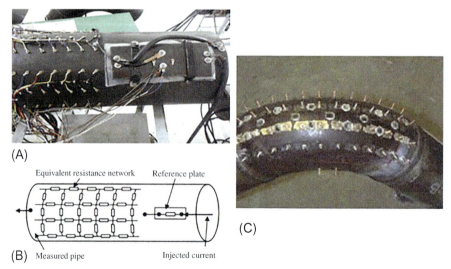

Fig. 10.10 Field signature monitoring, pitting corrosion.
Reproduced with permission from F. Gan, Z. Wan, Y. Li, J. Liao, W. Li, Improved formula for localized corrosion using field signature method, Measurement 63 (2015) 137–142, Copyright © 2014 Elsevier Ltd., Fig. 1.

10.4 Conclusions

The comparison of monitoring techniques has been done by oil and gas operators, Hedges et al. [96] have published in 2004 a comparative study of the different methods available on this time.

Sacrificial sand probes have a relatively simple design, they are exclusively intrusive, therefore, they can disturb the hydraulic flow, they can only be installed on accessible structures. The only indication that they transmit corresponds to the piercing of the membrane, the rate of metal loss can only be determined posteriori, and it is not very precise. The CAPEX is relatively low whereas the OPEX is potentially significant, essentially linked to the replacement of the probes and the maintenance operations related to this replacement.

Acoustic sand probes are nonintrusive, according to their design they can be installed on underwater structures, up to a depth of about 3000 m, they can detect particles of the size of the order of 15–25 µm inflows with a bulk velocity of 1 m s^{-1} minimum. The segregation between the noise due to the flow and the noise due to impact the particles is sometimes difficult. Welds penetrations can disturb the measurement. The CAPEX is more important that for sacrificial probes. The OPEX is mainly related to the maintenance operations of the connectors and associated instruments, the probes being fixed by clamps or collar, disassembly is easy, underwater interventions by ROV are possible.

The probes based on the measurement of the electrical resistance are all intrusive, only the flush type probes do not disturb too much the flow, the high-resolution probes are very sensitive due to the quality of their associated electronics and temperature compensation. Their lifespan is limited by the thickness of the sensitive element (generally half the thickness of the sensitive element available for metal loss). The CAPEX engaged is quite high, given the associated instrumentation and the data processing. The OPEX is also quite high. Some models can be installed in deep water, in this case, the thickness of the measuring element allows a service life of 25–30 years in theory.

The probes using ultrasound or EMAT technology are nonintrusive, they are installed on the surface, some models using the principle of surface tomography are designed to operate up to 3000-m deep. Conventional probes only detect a local indication, systems using tomography make it possible to assess the loss of material over an area of approximately 3 m^2.

The sensitivity of conventional UT or EMAT probes reaches 10 µm, that of tomography is around 0.1% of the wall thickness. CAPEX and OPEX are quite important.

The FSM and EFM technologies require a very large CAPEX, which is often difficult to justify in regard to the comparison of benefits provided by the other techniques available. OPEX is also very important, it is often linked to the instrumentation and associated connectivity, the interpretation of FSM and EFM measurements requires the intervention of specialists, mostly external to the production sites.

Monitoring development remains very important to be able to assess each of the erosion and corrosion components in the pipelines, devices have been adapted to carry out this type of study. Specific annular electrodes are used, the geometry of the test loops allows the flow pattern to be modified according to the tests. The total weight loss of the electrodes in the erosion-corrosion test is obtained from gravimetric measurements. The pure corrosion rate and the pure erosion rate are measured in a corrosive electrolyte free of abrasive particles and a noncorrosive suspension, respectively. Liu et al. [70, 91, 97] made a type of sensor that combines ER and LPR measurements, which allow the total loss of metal and the loss of metal by corrosion to be obtained simultaneously.

Finally, the positioning of probes is particularly important because corrosion and erosion are generally localized. As a matter of fact, the all-or-nothing sensors intended to evaluate the erosion component will be positioned upstream of an elbow, in the direction of circulation of the fluid while the sensors in acoustic or electrical systems will be positioned in the areas most subject to erosion and corrosion as shown in Fig. 10.11.

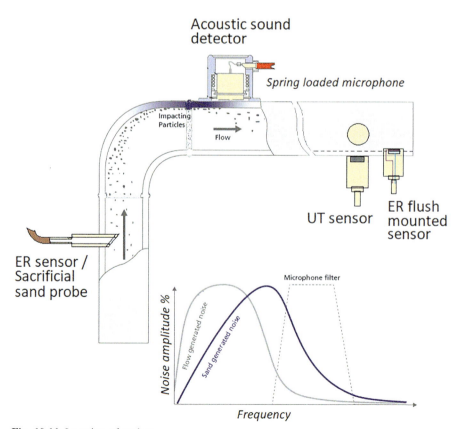

Fig. 10.11 Location of probes.

10.5 Notations

C	API 14E empirical constant	$[\text{kg}/(\text{m s}^{-2})]^{0.5}$	$[\text{lb}/(\text{ft s}^{-2})]^{0.5}$
D_i	Internal diameter of pipe	m	in.
d_{RCE}	Diameter of rotating cylinder electrode	cm	in.
E	Erosion rate	kg kg^{-1}	lb lb^{-1}
E_m	Erosion rate	kg s^{-1}	lb s^{-1}
E_k	Kinetic energy	J	
ER	Erosion rate	mm kg^{-1}	mil lb^{-1}
$F(\alpha)$	Function characterizing ductility of material		
F_M	Hardness parameter		
F_S	Particle (sand) angular parameter		
K	Material erosion constant	$(\text{m s}^{-1})^{-n}$	
m	Mass of impacting particle	kg	
m_p	Mass flow rate of particle	kg s^{-1}	lb s^{-1}
n	Material dependent index		
P	Hardness parameter	bar	
Re	Reynolds number		
r_{RCE}	Radius of RCE	cm	
R_{ref}^0	Initial electrical resistance of reference	Ω	
R_{ref}^t	Actual electrical resistance of reference	Ω	
R_{sen}^0	Initial electrical resistance of sensitive	Ω	
R_{sen}^t	Actual electrical resistance of sensitive	Ω	
S_k	Geometry-dependent constant		
S_m	Geometry-dependent constant		
t_k	Thickness of the detection element	mm	mils
U_m	Fluid (mixture) velocity	m s^{-1}	ft s^{-1}
U_p	Particle impact velocity	m s^{-1}	ft s^{-1}
V_e	Erosional velocity (API 14E)	m s^{-1}	ft s^{-1}
W	Sand production rate	kg day^{-1}	lb day^{-1}
x	Corrosion depth	μm	mils
ν	Kinematic viscosity of fluid	cm^2 s^{-1}	
ρ_m	Mixture density	kg m^{-3}	lb ft^{-3}
ω	Rotation speed	rad s^{-1}	

References

[1] G. Kock, J. Varney, N. Thompson, O. Moghissi, M. Gould, J. Payer, NACE IMPACT, International Measures of Prevention, Application, and Economics of Corrosion Technology Study, 2016. ©2016 NACE International.
[2] DNVGL-RP-O501: 2015, Managing Sand Production and Erosion.
[3] Energy Institute, Guidelines on Sand Erosion and Erosion-Corrosion Management, Energy Institute, June 2017, ISBN: 978 0 85293 998 7. Copyright © 2017 by the Energy Institute, London.
[4] G.I. Ogundele, W.E. White, Some observations on corrosion of carbon steel in aqueous environments containing carbon dioxide, Corrosion 42 (2) (1986) 71–78.

[5] J. Feyerl, G. Mori, S. Holzleitner, J. Haberi, M. Oberndorfer, W. Havlik, C. Monetti, Erosion-corrosion of carbon steels in a laboratory: three-phase flow, Corrosion 64 (2) (2008). Houston, TX.
[6] S. Tandon, M. Gao, R. McNealy, Erosion-corrosion failure of a carbon steel pipe elbow—a case study, in: Paper 09479, NACE Corrosion Conference & Expo, 2009. ©2009 NACE International.
[7] W.C. Krumbein, L.L. Sloss, Stratigraphy and Sedimentation, second ed., W.H. Freeman & Co Ltd, 1963. 1 janvier. ISBN-10: 0716702193, ISBN-13: 978-0716702191.
[8] A.M. Sherik, S.R. Zaidi, E.V. Tuzan, J.P. Perez, Black powder in gas transmission systems, in: NACE Corrosion 2008, Paper 08415, 2008.
[9] J. Yamada, K. Nakayama, H. Kaneta, Analyses of black powder in natural gas pipeline, in: NACE Corrosion 2011, Paper 11088, 2011.
[10] E. Zhang, D. Zeng, H. Zhu, S. Li, D. Chen, J. Lie, Y. Ding, G. Tian, Numerical simulation for erosion effects of three phase flow containing sulfur particles on elbows in high sour gas fields, Petroleum 4 (2018) 158–167.
[11] ANSI/NACE MR0175/ISO 15156-2015-SG, Petroleum, and Natural Gas Industries—Materials for Use in H_2S-Containing Environments in Oil and Gas Production.
[12] M. Aminul Islam, Z.N. Farhat, Mechanical and electrochemical synergism of API X42 pipeline steel during erosion–corrosion, J. Bio. Tribo. Corros. 1 (26) (2015).
[13] P.C. Okonowo, A.M.A. Mohamed, Erosion-corrosion in oil and gas industry: a review, Int. J. Metall. Mater. Sci. Eng. 4 (3) (2014) 7–28. ISSN(P): 2278-2516; ISSN(E): 2278-2524. © TJPRC Pvt. Ltd.
[14] C. De Waard, D.E. Milliams, Carbonic acid corrosion of steel, Corrosion 31 (5) (1975) 177–181.
[15] A. Kahyarian, B. Brown, S. Nešić, Mechanism of CO_2 corrosion of mild steel: a new narrative, in: Paper 11232, NACE Corrosion Conference & Expo, 2018. © 2018, NACE.
[16] H. Bai, Y. Wang, Y. Ma, Q. Zhang, N. Zhang, Effect of CO_2 partial pressure on the corrosion behavior of J55 carbon steel in 30% crude oil/brine mixture, Materials 11 (2018) 1765.
[17] J. Owen, C. Ramsey, R. Barker, A. Neville, Erosion-corrosion interactions of X65 carbon steel in aqueous CO_2 environments, Wear 0043-1648, 414–415 (2018) 376–389.
[18] L. Smith, K. de Waard, Corrosion prediction and materials selection for oil and gas environments, in: Paper No 05648, NACE Corrosion 2005, 2005.
[19] M.B. Kermani, A. Morshed, Carbon dioxide corrosion in oil and gas production—a compendium, Corrosion 59 (8) (2003). © 2003, NACE International.
[20] K.E. Heusler, Encyclopedia of Electrochemistry of the Elements, vol. 9, Marcel Dekker, New York, 1982.
[21] J.O. Bockris, D. Drazic, A.R. Despic, The electrode kinetics of the deposition and dissolution of iron, Electrochim. Acta 4 (2–4) (1961) 325–361.
[22] M. Keddam, O.R. Mattos, H. Takenout, Reaction model for iron dissolution studied by electrode impedance I. Experimental results and reaction model, J. Electrochem. Soc. 128 (1981) 257–266.
[23] H. Arabnejad, A. Mansouri, S.A. Shirazi, B.S. McLaury, Evaluation of Solid Particle Erosion Equations and Models for Oil and Gas Industry Applications, SPE-174987-MS, 2015. Copyright 2015, Society of Petroleum Engineers.
[24] S. Nešić, N. Thevenot, J.L. Crolet, D. Drazic, Electrochemical properties of iron dissolution in the presence of Co2—basics revisited, in: Corrosion, 1996. Paper No. 03.
[25] A. Kahyarian, M. Achour, S. Nešić, CO_2 corrosion of mild steel, in: Trends in Oil and Gas Corrosion Research and Technologies, 2017. Copyright © 2017 Elsevier Ltd.

[26] B.S. McLaury, J. Wang, S.A. Shirazi, J.R. Shadley, E.F. Rybicki, Solid particle erosion in long radius elbows and straight pipes, in: SPE Paper 38842, SPE Annual Technical Conference and Exhibition, II Production Operations and Engineering/General, San Antonio, Texas, October 1997.
[27] P.C. Okonkwo, M.H. Sliem, S.K. Mobbassar Hassan, R.A. Shakoor, A.M. Amer Mohamed, A.M. Abdullah, R. Kahraman, Erosion behavior of API X120 steel: effect of particle speed and impact angle, Coatings 8 (2018) 343.
[28] C. Kang, P.P. More, J.R. Vera, P. Altoe Ferreira, E. Costa Bastos, Flow patterns characteristics in multiphase riser pipelines, in: Paper 06589, 2006, Corrosion NACEexpo, 2006. © 2006 NACE International.
[29] H. Liu, W. Yang, R. Kang, A correlation for sand erosion prediction in annular flow considering the effect of liquid dynamic viscosity, Wear 404–405 (2018) 1–11.
[30] R. Rudman, H.M. Blackburn, The effect of shear thinning behavior on turbulent pipe flow, in: Third International Conference on CFD in the Minerals and Process Industries, CSIRO, Melbourne, Australia, 10–12 December, 2003. Copyright © 2003 CSIRO Australia.
[31] E. Volent, O.G. Dahlhaug, Literature review relevant to particle erosion in complex geometries, J. Phys. Conf. Ser. 1042 (2018).
[32] M.J. Bell, M. MacLeod, Assessment of particle erosion in coriolis meters, in: 30th International North Sea Flow Measurement Workshop 23–26 October 2012, La Houille Blanche/N° 7/8-192, 2012.
[33] M.E. Mohyaldin, N. Elkhatib, M. Che Ismail, Evaluation of different modelling methods used for erosion prediction, in: Paper 11005, NACE Shanghai China Section, 2011.
[34] M. Sukry Azizi Bin Wardi, The Impact of Sand Erosion in Gas Export Pipelines, Universiti Teknologi PETRONAS, 2015.
[35] P.G. Kumar, B.R.J. Smith, D. Vedapuri, J. Hariprasad, J. Subramani, L.D. Rhyne, Sand fines erosion in gas pipelines—experiments and CFD modeling, in: NACE-2014-3964, NACE International, Corrosion 2014, 2014.
[36] API RP 17B-5th edition, Recommended Practice for Flexible Pipe, 2014.
[37] A. Verheyleweghen, J. Jäschke, Oil Production optimization of several wells subject to choke degradation, IFAC-PapersOnLine 2405-8963, 51 (8) (2018) 1–6.
[38] K. Haugen, O. Kvernvold, A. Ronold, R. Sandberg, Sand erosion of wear-resistant materials: erosion in choke valves, Wear 186–187 (Part 1) (1995) 179–188.
[39] H.N. Raghavendra, M. Shivashankar, P.A. Ramalingam, Simulation of erosion wear in choke valves using CFD, Int. J. Eng. Res. Technol. 3 (7) (2014).
[40] B.F.M. Pots, J.F. Hollenberg, E.L.J.A. Hendriken, what are the real influences on flow on corrosion? in: Paper No. 06591, 2006, Corrosion NACExpo, 2006. © 2006 NACE International.
[41] M. Salasi, Synergism Between Abrasion and Corrosion, School of Mechanical and Chemical Engineering, University of Western Australia, 2012.
[42] R.J.K. Wood, Erosion-corrosion synergism for multi-phase flowline materials, La Houille Blanche 7–8 (December 1992) 605–610.
[43] Y. Xu, J.-L. Luo, M.Y. Tan, An overview of techniques for measuring the interaction between erosion and corrosion, in: Corrosion & Prevention 2017 Paper 99, 2017, pp. 1–12.
[44] H.A. Kishawya, H.A. Gabbar, Review of pipeline integrity management practices, Int. J. Press. Vessel. Pip. 87 (2010) (2010) 373–380.
[45] API RP 14E-5th edition, Reaffirmed 2019, Recommended Practice for Design and Installation of Offshore Products Platform Piping Systems.

[46] ISO 10423:2009, confirmed 2015, Petroleum and natural gas industries—Drilling and production equipment—Wellhead and christmas tree equipment.
[47] ISO 13703:2000, confirmed 2015, Design and installation of piping systems on offshore production platforms.
[48] N.A. Barton, Erosion in Elbows in Hydrocarbon Production Systems: Review Document, Prepared by TÜV REL Limited for the Health and Safety Executive, Research Report 115, 2003.
[49] A. Moncur, A. Kinsler, Optimising Production Through Sand Management: A Case Study from the United Kingdom Continental Shelf (UKCS), Sand Monitoring Services Ltd, n.d.
[50] C. McPhee, C. Webster, G. Daniels, C. Reed, F. Mulders, C. Howat, A. Britton, Developing an integrated sand management strategy for Kinabalu field, offshore Malaysia, in: SPE 170749, 2014. Copyright 2014, Society of Petroleum Engineers.
[51] F. Dehghani, Sand production management for oil and gas reservoirs, in: F. Kongoli, M. Gaune-Escard, T. Turna, M. Mauntz, H.L. Dodds (Eds.), Sustainable Industrial Processing Summit SIPS 2016. Volume 9: Molten Salts and Ionic Liquids, Energy Production, vol. 9, FLOGEN Star Outreach, Montreal (Canada), 2016, pp. 363–370.
[52] J.W. Martin, Y. Sun, J. Alvarez, E. Babaian-Kibala, S. Hernandez, Design and operations guidelines to avoid erosion problems in oil and gas production systems—one operator's approach, in: Paper 06592, Corrosion NACEexpo 2006, 2006. ©2006 NACE International.
[53] S. Whitfield, Managing the Effects of Sand Erosion, Oil Gas Facilities, 28 February 2017. *https://pubs.spe.org/en/ogf/ogf-article-detail/?art=2769*.
[54] Y. Loong, Murphy, H. Rawlins, Upgrade of spar topsides with comprehensive facilities sand management system, in: OTC-24705-MS, Offshore Technology Conference, 2014. Copyright 2014.
[55] D. Clarke, Pipeline integrity monitoring: developments in non-intrusive flow-through devices, in: Pipeline Technology Conference 2009, 2009.
[56] NORSOK Standard M-001:2004, Materials selection.
[57] ISO 21457: 2010, confirmed 2018, Petroleum, petrochemical and natural gas industries— Materials selection and corrosion control for oil and gas production systems.
[58] B.S. Mann, V. Arya, A.K. Maiti, M.U.B. Rao, P. Joshi, Corrosion and erosion performance of HVOF/TiAlN PVD coatings and candidate materials for high pressure gate valve application, Wear 260 (2006) 75–82. © 2005 Elsevier B.V.
[59] H.E. Rincon, Testing and Prediction of Erosion-Corrosion for Corrosion Resistant Alloys Used in the Oil and Gas Production Industry, The University of Tulsa, 2006.
[60] R.J.K. Wood, Erosion corrosion Interactions and their effect on marine and offshore components, in: Eurocorr 2004, 2004.
[61] B.J. Usman, S.A. Ali, Carbon dioxide corrosion inhibitors: a review, Arab. J. Sci. Eng. 43 (2017) 1–22.
[62] C.M. Canto Maya, Effect of Wall Shear Stress on Corrosion Inhibitor Film Performance, Russ College of Engineering and Technology of Ohio University, December 2015. © 2015 Christian M. Canto Maya.
[63] O.O. Ige, R. Barker, X. Hua, L.E. Umoru, A. Neville, Assessing the influence of shear stress and particle impingement on inhibitor efficiency through the application of in-situ electrochemistry in a CO_2-saturated environment, Wear 304 (2013) 49–59.
[64] Y.Z. Xu, et al., Corrosion behavior of pipeline steel under deposit corrosion and the inhibition performance of organic phosphine inhibitor, Acta Metall. Sin. (Chin. Ed.) 52 (2016) 320–330.

[65] Y. Ding, Mechanistic Understanding of CO_2 Corrosion Inhibition at Elevated Temperatures, Russ College of Engineering and Technology of Ohio University, May 2019. © 2019 Yuan Ding.
[66] A. Berradja, Singh electrochemical techniques for corrosion and tribocorrosion monitoring: fundamentals of electrolytic corrosion, in: Corrosion Inhibitors, IntechOpen, 2019, https://doi.org/10.5772/intechopen.85392 (Chapter 9).
[67] S. Ramachandran, Selection of Corrosion Monitoring Equipment for Subsea Safety Joint (Master thesis), University of Stavanger, 2016.
[68] K. Alawadhi, A.S. Aloraier, S. Joshi, J. Alsarraf, S. Swilem, Investigation on preferential corrosion of welded carbon steel under flowing conditions by EIS, J. Mater. Eng. Perform (2013), https://doi.org/10.1007//s11665-013-0525-z. © ASM International.
[69] M.H. Buszko, A.K. Krella, Slurry erosion—design of test devices, Adv. Mater. Sci. 17 (2) (2017). 52.
[70] Y. Xu, L. Liu, Q. Zhou, X. Wang, M.Y. Tan, Y. Huang, An overview of major experimental methods and apparatus for measuring and investigating erosion-corrosion of ferrous-based steels, Metals 10 (2020) 180.
[71] Y. Xu, L. Liu, Q. Zhou, X. Wang, Y. Huang, Understanding the influences of pre-corrosion on the erosion-corrosion performance of pipeline steel, Wear 442–443 (2020).
[72] H. Arabnejad, A. Mansouri, S.A. Shirazi, B.S. McLaury, Development of mechanistic erosion equation for solid particles, Wear 0043-1648, 332–333 (2015) 1044–1050.
[73] A. Abdulla, Estimating Erosion in Oil and Gas Pipeline Due to Sand Presence, Department of Mechanical Engineering Blekinge Institute of Technology, Karlskrona, 2011. Independent thesis advanced level.
[74] F.M. Sani, S. Nesic, S. Huizinga, K. Esaklul, Review of the API RP 14E erosional velocity equation: origin, applications, misuses and limitations, in: Paper 13206, NACE Corrosion Conference & Expo 2019, 2019. ©2019 NACE International.
[75] H. Mansouri, Applying higher C-values in API RP 14E erosion velocity calculations for gas condensate wells, in: Conference Paper, 51318-10627 SG—A Case Study NACE, 2018.
[76] M.M. Salama, An alternative to API 14E erosional velocity limits for sand laden fluids, in: Offshore Technology Conference, 4–7 May, Houston, Texas, 1998.
[77] S.A. Shirazi, B.S. McLaury, J.R. Shadley, E.F. Rybicki, Generalization of the API RP 14E guideline for erosive services, SPE28518, J. Petrol. Technol. 1995 (1995) 693–698.
[78] A. Huser, O. Kvernvold, Prediction of sand erosion in process and pipe components, in: Proc. 1st North American Conference on Multiphase Technology, Banff, Canada, 1998, pp. 217–227.
[79] M.M. Salama, E.S. Venkatesh, Evaluation of API RP 14E erosional velocity limitation for offshore gas wells, in: OTC 4485, Offshore Technology Conference, Houston, Texas, 1983.
[80] S.J. Svedeman, K.E. Arnold, Criteria for sizing multiphase flow lines for erosive/corrosive service, in: SPE 26569, 68th Annual Technical Conference of the Society of Petroleum Engineers, Houston, Texas, 1993.
[81] A.T. Bourgoyne, Experimental study of erosion in diverter systems, in: SPE/IADC 18716, Proc SPE/IADC Drilling Conference, New Orleans, 28 February–3 March, 1989, pp. 807–816.
[82] M.M. Salama, An alternative to API 14E erosional velocity limits for sand laden fluids, in: OTC 8898, Offshore Technology Conference, Houston, Texas, 1998, pp. 721–733.

[83] P.D. Weiner, G.C. Tolle, Detection and Prevention of Sand Erosion of Production Equipment, API OSAPR Project No 2, Research Report, Texas A&M University, College Station, TX, 1976.

[84] J.K. Edwards, B.S. McLaury, S.A. Shirazi, Evaluation of alternative pipe bend fittings in erosive service, in: Proceedings of ASME FEDSM'00: ASME 2000 Fluids Engineering Division Summer Meeting, Boston, June 2000.

[85] S.A. Shirazi, B.S. McLaury, H. Arabnejad, A semi-mechanistic model for predicting sand erosion threshold velocities in gas and multiphase flow production, in: Conference Paper, SPE-181487-MS, September 2016.

[86] J.K. Edwards, B.S. McLaury, S.A. Shirazi, Supplementing a CFD code with erosion prediction capabilities, in: Proceedings of ASME FEDSM'98: ASME 1998 Fluids Engineering Division Summer Meeting, Washington DC, June 1998.

[87] M.S. Wallace, CFD Based Erosion Modelling of Simple and Complex Geometries (Ph.D. thesis), University of Strathclyde, Glasgow, 2001.

[88] P.M. Birchenough, S.G.B. Dawson, T.J. Lockett, McCarthy, Simultaneous erosion and corrosion in multi-phase flow, in: NACE 7th Middle East Conference on Corrosion, Bahrain, 1996.

[89] CEFRACOR, Commission Corrosion dans les industries pétrolières, gazières et chimiques, Monitoring de la corrosion interne dans les industries pétrolières et gazières, EDP Sciences, 2016, ISBN: 978-2-7598-1746-7.

[90] K. Wold, R. Stoen, M. Rapone, Integration of intrusive and non-intrusive methods for corrosion and sand/erosion monitoring, La Metallurgia Italiana (2) (2012).

[91] L. Liu, Y. Xu, C. Xu, X. Wang, Y. Huang, Detecting and monitoring erosion-corrosion using ring pair electrical resistance sensor in conjunction with electrochemical measurements, Wear 428 (2019) 328–339. 0043-1648/ © 2019 Published by Elsevier B.V.

[92] F. Simonetti, P.B. Nagy, G. Instanes, A.O. Pedersen, Ultrasonic computerized tomography for continuous monitoring of corrosion and erosion damage in pipelines, in: NACE-2015-5750 Paper 50212, 2015, Corrosion 2015, 15-19 March, Dallas, Texas, 2015.

[93] F. Gan, Z. Wan, Y. Li, J. Liao, W. Li, Improved formula for localized corrosion using field signature method, Measurement 63 (2015) 137–142. © 2014 Elsevier Ltd.

[94] M. Ho, S. El-Borgi, D. Patil, G. Song, Inspection and monitoring systems subsea pipelines: a review paper, Struct. Health Monit. 19 (2) (2020) 606–645. © The Author(s) 2019.

[95] R. Johnsen, B.E. Bjornsen, D. Morton, D. Parr, B. Ridd, Weld root corrosion monitoring with a new electrical field signature mapping inspection tool, in: NACE Corrosion, 2000, Paper 00096, 2000.

[96] B. Hedges, A. Bodington, A comparison of monitoring techniques for improved erosion control, a field study, in: NACE Conference Paper, 2004/01/01, 2004. Product Number: 51300-04355-SG.

[97] Y.-Z. Xu, Y. Huang, W. Xiaona, X. Lin, Experimental study on pipeline internal corrosion based on a new kind of electrical resistance sensor, Sens. Actuators B 224 (2016) 37–47, https://doi.org/10.1016/j.snb.2015.10.030.

Index

Note: Page numbers followed by *f* indicate figures and *t* indicate tables.

A

Acoustic-based nonintrusive sand monitoring sensors, 248–249, 251
Analysis of variance (ANOVA), 87–88
Archard and friction energy wear approach, 138–141
Archard's law, 29, 159
Atomic force microscopy (AFM), 40, 49–54
Attenuated total reflectance (ATR)-FTIR, 55

B

Barite, 229
Black powder, 229
Bowden and Tabor theory, 146–147

C

Calcite, 229
Carbon dioxide (CO_2) corrosion, 230
Circularly translating pin-on-disk (CTPOD) tester, 91, 92*f*, 94, 95*f*, 101–102, 103*f*
Clearance, 78
Coating durability, 158–161
CoCrMo-alloy bioimplant, 54
Coefficient of restitution, 193–195
Compressive stresses, 8–12
Computational fluid dynamics (CFD), 244
Contact oxygenation concept (COC), 143–145, 161–162
Contact stresses, 68, 74–75
Correlation, 112. *See also* Pearson correlation method (PCM)
Corrosion/formation products, 229
Corrosion-resistant alloys (CRAs), 239
Corrosive materials, 230
Cracks, 18–19
Crevice effects, 18–19
Cup positioning, 77

D

Dimensionless erosion *vs.* impact angle graph, 200–201, 201*f*
Direct impact model (DIM), 243–244
Dislocations effects, 20
Dissolution-repassivation, reaction model of, 180–182, 186, 187*f*
Drill cuttings, 229
Dual mobility, 67, 72–75

E

Electrical field method (EFM), 250
Electrochemical impedance spectroscopy (EIS), 32, 43, 179–186
Electrochemical microcell (EMC), 37–38
Electrochemical techniques, 111
Electron backscatter diffraction (EBSD), 16, 20, 22–23
Electron work function (EWF), 11–12
Erosion-corrosion, 227
  laboratory and loop tests, 239–241
  management strategy, 231–250
  models, 242–244
  oil and gas production, 228–231
  risk assessment, 235–236
  on site monitoring, 244–250
  system and subsystem components, 230–231
Erosion-corrosion maps
  boundary conditions, 199–200
  corrosion rates, 198–199
  elastic rebound effects, 200–222
  particle rebound effects, 195–198
  velocity-impact angle maps, 211–223
  velocity-potential regime maps, 200–211, 222–223
Erosion models, 195–198
Erosive agents, 229
Erosive service, 235
European Corrosion Federation (EFC), 1

F

Falex Miller tester, 90–91, 91f
Falex tribology 10-station wear tester, 91–93, 93f
Faraday's law, 29
Fatigue corrosion (FC), 123–124
Fatigue cracks, 131
Field signature method (FSM) method, 250
Fretting mapping approach, 137
Fretting wear analysis
  ambient conditions
    aqueous processes, 153–154
    lubricated interfaces, 152–153
    temperature, 150–152
  contact loadings
    contact size, 148–149
    extended wear coefficient approach, 149–150
    normal load, 146–147
    sliding amplitude, 145–146
    sliding frequency, 147–148
  experiments, 132–137
Friction energy density parameter, 158–161
Friction energy wear approach, 138–141
FTIR-spectroscopy, 54–55

G

Geometrical effects, 16–18
Geometrically necessary dislocations (GND), 16, 20
Grain refinement, 20–23
Gross slip regime (GSR), 135–137

H

Hertz's theory, 67
Hip implant prosthesis, 67, 68f

I

Immersion tests, 122
In situ vibrational spectroscopy, 54–58

L

Lab testing approach, 83–85
Laser shock processing (LSP), 8, 11, 11f
Local electrochemical impedance spectroscopy (LEIS), 43–46, 60
Local electrochemical techniques, 34–54
Lubricated interfaces, 152–153

M

Material degradation, 1, 2f
Material response fretting maps (MRFM), 137
Materials performance maps, 221f, 222
Maximum wear depth, 154–161
Mechanical loading, 1
Mechanical stresses, 3
Microtribocorrosion measurements, 37–38, 37f
Miller number (MN), 95–97
Mindlin's theory, 134–135
Mixed fretting regime (MFR), 135–137
Mott-Schottky analysis, 43

N

Nano-IR spectroscopy, 58, 60f
Ni-P nanocomposites, frictional behavior of, 106, 106f
Nonerosive agents, 229
Nonintrusive probes, 248
Nyquist plots, 183–185, 184f, 186f

O

Oil and gas, 227–231
One-way ANOVA analysis, 88–89, 89f
On site erosion-corrosion monitoring, 244–250

P

Parallel wear tests, 89
Partial slip regime (PSR), 135–137
Particle impact velocity-impact angle wastage maps, 207–210f, 211, 217–220f, 222
Passive film modifications, 19
Pearson correlation method (PCM)
  application
    corrosion science and engineering, 120–122
    fatigue corrosion, 123–124
    stress corrosion cracking, 123–124
    tribocorrosion, 124–127
  assumptions, 112–113
  basic principles, 113–114
  methodology, 119
  output, 116–118
  significance test, 114–116
  variables, 111–112

Index

Photothermal-induced resonance (PTIR), 58
Piezoelectric ultrasonic thickness gauges, 249
Plastic deformation, 15–23
Pressure-mapping film, 78–79
Pressure-sensitive film approach, 68, 71, 74–76
Printing roller, confocal analysis on, 97–98, 99f
"Problem tree" analysis, 5f, 6
Produced sand, 229
Proppants, 229

R

Raman spectroscopy, 54–55
Rate of erosion, 197
Repassivation kinetics, 173–178
Ring pair resistance electrical resistance sensors (RPERS), 248
Rotating disc electrode (RDE) test, 240
Running condition fretting map (RCFM), 137

S

Sacrificial sand probes, 246, 251
Salt fog atmosphere tests, 122
Sand production
  control strategy, 237
  corrosion inhibitors, 239
  management strategy, 236–237, 237f
  material selection, 239
  mitigation measures, 238–239
  selection criteria and methodology, 238
Scanning electrochemical microscopy (SECM), 46–49, 60
Scanning Kelvin probe (SKP), 38–43, 60
Scanning Kelvin probe force microscopy (SKPFM), 49–54
Scanning reference electrode technique (SRET), 34–37, 60
Scanning vibrating electrode technique (SVET), 34–37, 60, 111
Scattering scanning near field optical microscopy (sSNOM), 58
Sensitive pressure Fujifilm-prescale, 68–70, 69f
Severe plastic deformation (SPD), 20–23
Shales, 229
Significance test, 114–116
Simplest erosion model, 242

Slurry abrasivity response (SAR) number, 95–97
Slurry jet impingement rig systems, 241
Solid particle erosion event, energy balance during, 192–193, 192t
Standard electrical resistance (ER) probes, 247
Statistically stored dislocations (SSD), 16, 20
Statistical wear analysis
  aeronautical field, 101–102
  biomedical field, 94–95
  construction industry, 95–97
  industrial coatings, 106–107
  need for, 85–89, 94–107
  offshore installations, 102–105
  printing industry, 97–100
Stress corrosion cracking (SCC), 123–124
Surface wear modeling, 154–161

T

Tekscan sensors, 70, 71–72f, 72–73
Tensile stresses, 12–15
Test flow loops, 241
Third-body theory (TBT), 141–143, 161–162
Tolerable erosion and corrosion, 233–234
Triboelectrochemical admittance, 174, 176
Triboelectrochemical impedance, 172, 176
Tukey tests, 87–89

V

Variance, 83
Velocity-impact angle maps, 211–223
Velocity-potential regime maps, 200–211, 222–223
Vibrational spectroscopy, 54–58, 61
Vibrational sum frequency spectroscopy (VSFS), 55–58
Volta potentials, 38–39, 61
Volume of unrelaxed crater, 194

W

Wear rate, 138–145
Wear simulation, 84–85, 154–158
Wear volume, 138–141, 152–154, 158
Wire beam electrode (WBE) technique, 241
Worn profiles modeling, 154–158
Worn surfaces, 30, 30f

Printed in the United States
by Baker & Taylor Publisher Services